CROSSING THE RIVER BY TOUCHING THE STONES
ALTERNATIVE APPROACHES IN TECHNICAL AND VOCATIONAL EDUCATION AND TRAINING IN THE PEOPLE'S REPUBLIC OF CHINA AND THE REPUBLIC OF KOREA

JANUARY 2022

ASIAN DEVELOPMENT BANK

 Creative Commons Attribution 3.0 IGO license (CC BY 3.0 IGO)

© 2022 Asian Development Bank
6 ADB Avenue, Mandaluyong City, 1550 Metro Manila, Philippines
Tel +63 2 8632 4444; Fax +63 2 8636 2444
www.adb.org

Some rights reserved. Published in 2022.

ISBN 978-92-9269-349-7 (print); 978-92-9269-350-3 (electronic); 978-92-9269-351-0 (ebook)
Publication Stock No. TCS220019-2
DOI: http://dx.doi.org/10.22617/TCS220019-2

The views expressed in this publication are those of the authors and do not necessarily reflect the views and policies of the Asian Development Bank (ADB) or its Board of Governors or the governments they represent.

ADB does not guarantee the accuracy of the data included in this publication and accepts no responsibility for any consequence of their use. The mention of specific companies or products of manufacturers does not imply that they are endorsed or recommended by ADB in preference to others of a similar nature that are not mentioned.

By making any designation of or reference to a particular territory or geographic area, or by using the term "country" in this document, ADB does not intend to make any judgments as to the legal or other status of any territory or area.

This work is available under the Creative Commons Attribution 3.0 IGO license (CC BY 3.0 IGO) https://creativecommons.org/licenses/by/3.0/igo/. By using the content of this publication, you agree to be bound by the terms of this license. For attribution, translations, adaptations, and permissions, please read the provisions and terms of use at https://www.adb.org/terms-use#openaccess.

This CC license does not apply to non-ADB copyright materials in this publication. If the material is attributed to another source, please contact the copyright owner or publisher of that source for permission to reproduce it. ADB cannot be held liable for any claims that arise as a result of your use of the material.

Please contact pubsmarketing@adb.org if you have questions or comments with respect to content, or if you wish to obtain copyright permission for your intended use that does not fall within these terms, or for permission to use the ADB logo.

Corrigenda to ADB publications may be found at http://www.adb.org/publications/corrigenda.

Note:
In this publication, "$" refers to United States dollars.
ADB recognizes "China" as the People's Republic of China; "Korea" and "South Korea" as the Republic of Korea; "Russia" as the Russian Federation; "Britain" as the United Kingdom; and "Peking" as Beijing.

On the cover: (1) Batam State Polytechnic students learn how to work with metal at the welding and plasma cutting lab. This subject is part of the school's newly introduced aircraft maintenance diploma course. (2) SMKN 2 Depok is also a partner of Toyota Astra Motor. A class of 32 students now studies in a large training center filled with almost exactly the same equipment used in actual Toyota plants, including a modern automotive body-painting chamber provided by Toyota. Most, if not all, of the students will be hired by the company. The rest will likely be hired by other automotive companies. (3 and 4) Students attending vocational schools in Karakol, Kyrgyz Republic (photos by ADB).

Cover design by Joe Mark Ganaban.

CONTENTS

Tables, Figures, and Boxes -- v
Foreword --- vi
Acknowledgments -- vii
Abbreviations --- viii
Executive Summary -- ix

1. **MODEL OF TECHNICAL AND VOCATIONAL EDUCATION AND TRAINING IN THE WORLD** -- 1

   1.1. Education for All versus Technical and Vocational Education and Training: Shifts in Priorities 2
   1.2. Trends in Technical and Vocational Education and Training Development— the Emergence of the "Building-Block" Approach 4
         National Vocational Qualifications Frameworks 6
         Competency-Based Curriculums 7
         Public–Private Partnerships 8
         Labor Management Information Systems 8
         Sector Skills Councils 9
   1.3. Typologies of Technical and Vocational Education and Training Systems in the Developed World 10
   1.4. Country Cases—Role of Culture in Technical and Vocational Education and Training Within Germany and the United Kingdom 12
         Phase 1: The Era of Guilds and Apprenticeships 12
         Phase 2: Industrial Revolution and Collapse of the Guilds 13
         Phase 3: Postindustrialization Expansion of Education 14
   1.5. Limited Successes in Transplanting Models of Technical and Vocational Education and Training 15
   1.6. Policy Implications and Conclusion 16
         On Borrowing 16
         On Importance of History and Culture 16

2. **EXPERIENCE OF THE REPUBLIC OF KOREA** — 18

   2.1. Identification of Labor Market Skill Needs When Sector Skills Councils Could Not Be Formed 20
   2.2. Economic Development through Aggressive Skills Policies Focusing on Future Skills Demand 23
   2.3. Status of Technical and Vocational Education and Training Policy Within the Whole Economic Development Strategy and the Conditions for Successful Integration 28
   2.4. Pursuit of Public–Private Partnerships and the Role of Government in Driving Such Cooperation 32
   2.5. Policies and Measures for Promoting Employer-Led Training 36
   2.6. Contributions of the Qualifications System Despite the Absence of a Comprehensive Qualifications Framework for All Education and Training Programs 39
   2.7. Relationship Between the Development of General Education and Technical and Vocational Education and Training 43

3. **EXPERIENCE OF THE PEOPLE'S REPUBLIC OF CHINA** — 49

   3.1. Identification of Labor Market Skill Needs When Sector Skills Councils Could Not Be Formed 52
   3.2. Economic Development through Aggressive Skills Policies Focusing on Future Skills Demand 56
   3.3. Status of Technical and Vocational Education and Training Policy Within the Whole Economic Development Strategy and the Conditions for Successful Integration 59
   3.4. Pursuit of Public–Private Partnerships and the Role of Government in Driving Such Cooperation 62
   3.5. Policies and Measures for Promoting Employer-Led Training 67
   3.6. Contributions of the Qualifications System Despite the Absence of a Comprehensive Qualifications Framework for All Education and Training Programs 72
   3.7. Relationship Between the Development of General Education and Technical and Vocational Education and Training 79

4. **CONCLUSION** — 83

   4.1 Lessons Learned 83
   4.2 Final Note 86

REFERENCES — 89

TABLES, FIGURES, AND BOXES

TABLES
1. World Bank Education Policy Swings Over Time 4
2. Building-Block Approaches for Technical and Vocational Education and Training 5
3. Key Areas in Technical and Vocational Education and Training Transfer 6
4. Upper Secondary Students by Type of Educational Program 47

FIGURES
1. Four Technical and Vocation Education and Training Regimes 10
2. Evolution of Two Technical and Vocational Education and Training Regimes 13
3. Enrollment Rate of Students by Level of Education 45
4. Integrating Vocational Education with National Development Strategy 61
5. Typology of Public–Private Partnership Coordination Model 64
6. Typology of Skills Partnership 69
7. People's Republic of China—National Qualifications System Since 2000 73
8. National and Regional Occupation Testing Authorities 75
9. Occupational Skills and Testing Authority Certification Participants and Pass Rates, 1996–2014 76
10. Vocational and General High School Enrollment Share, 1980–2018 80

BOXES
1. Sector Organization as an Industry Skills Engine 54
2. Role of Local Government and Industry Alliances 55
3. Bridging Industry Needs with Vocational Education and Training 58
4. Government Agency as Intermediary Agency 65
5. Business Associations as Public–Private Partnership Intermediaries 65
6. Parental Model in Strategic Industry 70
7. Shenzhen Polytechnic Co-Evolving with Huawei 71

v

FOREWORD

Economic diversification requires a critical mass of skilled technicians in respective fields, where technical and vocational education and training (TVET) plays an instrumental role. Before the coronavirus disease (COVID-19) pandemic, many Asian countries and regions, including South Asia, had shown remarkable economic growth. This economic diversification supported by a skilled workforce would be important for quick economic recovery and sustainable economic growth. Furthermore, during the pandemic, international travel restrictions were imposed, and the governments in respective countries realized that developing and maintaining a minimum number of skilled labor such as utility repair technicians and skilled construction workers is important from the national security point of view. Governments have high expectations on TVET for employment, economic recovery, and national security, but TVET faces challenges such as social stigma, shortage of skilled instructors, and insufficient industry partnership in developing Asia.

Knowledge transfer from successful TVET models in Asia could inspire policy makers to think outside the box. The building blocks of TVET are well known but mostly based upon successful cases in Europe, such as Germany's dual training programs. While there is a lot to learn from successful TVET history and practices in Europe, different TVET pathways exist in East Asia such as in the People's Republic of China (PRC) and the Republic of Korea. This publication is titled "Crossing the River by Touching the Stones," which is a metaphor for steadily exploring the next course of action that is often used by leaders in the PRC in describing paths for their economic reform. Many good practices are available, and policy makers are encouraged to look for suitable approaches before taking next policy reforms.

The East Asian good practices, however, are not necessarily well understood, and this publication aims to close the knowledge gap. In particular, this publication comes up with seven relevant policy questions based on common approaches taken for TVET reform, and which show alternative pathways taken by the PRC and the Republic of Korea. This publication reconfirms that there is no "one-size-fits-all" for TVET, and policy makers in Asia are expected to be motivated to explore different TVET models, and come up with new ways of TVET delivery that best suit their own country context.

Sungsup Ra
Chief Sector Officer
Sustainable Development and Climate Change Department
Asian Development Bank

ACKNOWLEDGMENTS

This technical study consists of four chapters. Chapter 1 (introduction) and Chapter 4 (conclusion) were led by Ha Wei, associate professor and associate dean, the Graduate School of Education, Peking University, with Conor Mccutcheon, research assistant, Graduate School of Education, Peking University. Chapter 2 (experience of the Republic of Korea) was written by Youngsup Choi, professor, Techno-HRD Graduate School at Korea University of Technology and Education, who has a long experience in technical and vocational education and training (TVET) research in the Republic of Korea. Chapter 3 (experience of the People's Republic of China [PRC]) was authored by Po Yang, associate professor, the Graduate School of Education, Peking University, based on her extensive research in the PRC. Sungsup Ra, chief sector officer, Sustainable Development and Climate Change Department, Asian Development Bank (ADB) provided overall guidance with the support of Ryotaro Hayashi, social sector economist, Human and Social Development Division, South Asia Department (SAHS), ADB. The peer reviewers were Fook Yen Chong, senior social sector specialist (skills development), Human and Social Development Division, Southeast Asia Department, ADB; and Chaoyi Hu, ADB consultant, who provided valuable comments from their project management and regional cooperation experiences related to TVET. Ma. Cristina S. Bardos of SAHS provided administrative assistance.

ABBREVIATIONS

ADB	–	Asian Development Bank
CNY	–	Chinese yuan
CPC	–	Communist Party of China
EPB	–	Economic Planning Board
ICT	–	information and communication technology
ILO	–	International Labour Organization
PPP	–	public–private partnership
PRC	–	People's Republic of China
SMEs	–	small and medium-sized enterprises
SOE	–	state-owned enterprise
TVET	–	technical and vocational education and training
UNESCO	–	United Nations Educational, Scientific and Cultural Organization
UNEVOC	–	International Centre for Technical and Vocational Education and Training
VET	–	vocational education and training

EXECUTIVE SUMMARY

This study titled *Crossing the River by Touching the Stones: Alternative Approaches in Technical and Vocational Education and Training* begins with a critical review of popular technical and vocational education and training (TVET) models for international development. Different types of TVET systems in Europe are introduced, exemplified by Germany and the United Kingdom (UK), to illustrate the diverse paths in developing a TVET system. These are juxtaposed against the East Asian experiences, as exemplified by the People's Republic of China (PRC) and the Republic of Korea. Through the case studies in the PRC and the Republic of Korea, this study aims to show different pathways for TVET delivery in different historical and cultural contexts.

Chapter 1 focuses on trends and approaches in TVET for international development. The chapter discusses the building-block approach based largely on European experiences such as the dual training system in Germany, which was adopted in 19 TVET projects of the Asian Development Bank (ADB) from 2010 to 2020. The chapter likewise looks into failures in replicating successful TVET models in Europe in developing countries due to differences in history and culture. While the German TVET system has gained popularity in international TVET development, institutional, cultural, and sociological mismatches and power differentials make its system transfer difficult. Despite genuine good intentions, without a full understanding of the historical and institutional environment in a particular locality, governments in developing countries might be tempted to emulate the foreign models in wholesale fashion, which has a higher chance of failure.

This sets the stage for the country TVET case studies on the Republic of Korea in Chapter 2 and the PRC in Chapter 3. These two East Asian countries have gone through rapid economic transformation in the last few decades, and their homegrown skills development experiences and solutions could be a good reference for other countries. While simply borrowing an entire system that has been shaped around a completely different historical and cultural context will rarely have the desired results, governments can still explore partial borrowing, pilot borrowing, customization, and conceptual borrowing to develop a unique TVET system that fits into their local context.

Chapters 2 and 3 address the following seven questions as a springboard for the East Asian case studies on the PRC and the Republic of Korea, from which key lessons emerge, as follows:

(i) *Sector skills councils.* On the role of sector skills councils in identifying the labor market's skills needs, how was demand measured when the industrial base and employer roles were very weak and sector skills councils could not be formed?

In the early stages of economic development, the government plays a key role in assessing the skills demand. Employer-led organizations such as sector skills councils cannot articulate the skills demand because employers only need low-wage unskilled workers to manage their business. In the Republic of Korea, the government set up a vocational training standards deliberation committee involving 200 experts who largely came from universities or technical high schools in the 1960s. In the PRC, under the guidance of the State Planning Commission and State Economic Commission, the Ministry of Labor led skills coordination and forecasting in 1950s, particularly for state-owned enterprises in urban areas. The local government also played an active role in assessing industry skills demand in the region.

(ii) *Skills development for the future.* When the industrial base was weak, how did the government pursue development of industry through aggressive skills development policies? In such cases, can TVET policy be said to take the leading role in future-oriented skills development, rather than a passive response to labor demand?

The two countries invested in school-based training but with different approaches. In the Republic of Korea, there was significant investment in selected technical high schools. An exemplary case is Kumoh Technical High School established in 1973. The best students were attracted with sophisticated infrastructure and well-trained teachers, and graduates became industry leaders driving economic transformation. This approach also helped address the social stigma attached to TVET. The PRC's investment in school-based training was relatively modest but it came up with successful buffer organizations called industry teaching and education steering committees starting in 1999 to connect schools and enterprises. The Republic of Korea's aggressive legislative discussions mandating employers to provide vocational training also changed the mindset of employers to develop a future-oriented skilled workforce by themselves.

(iii) *TVET policy in economic development.* What was the status of TVET policy in the economic development strategy under the strong leadership of the government? What were the conditions for the government to successfully integrate TVET policy into its economic development strategy?

The highest level of political leadership showed strong interest in promoting skills development in the early economic development stages of both countries. Capable bureaucrats translated this vision into actual TVET policy and sector implementation, which were closely aligned with national strategies such as 5-year economic development plans. In the Republic of Korea, effective advice and support from foreign countries from late 1960s to early 1980s were also instrumental in developing and implementing TVET policy.

(iv) *Public–private partnership.* For the government's skills development policy to be implemented effectively, enormous resources are needed, and public–private partnerships (PPPs) are critical to securing and enforcing them. How did this cooperation really happen? Did the public and private sectors have partnerships on equal footing? What was government's role in driving such cooperation? What incentives and regulations were applied? Under what conditions could such incentives and regulations work effectively?

The Korean government provided aggressive incentives to engage private TVET providers. Land reform in 1949 incentivized large landholders to allocate space for establishing private schools. Financial incentives such as subsidies on operating cost and tax exemptions, though not always granted, were also helpful in expanding access to and improving the quality of TVET, and ensuring proper use of public resources. In addition, deregulation in TVET supported the private sector in establishing and managing their training centers. In the PRC, different PPP models were utilized depending on the degree of interfirm cooperation and school–firm cooperation, with intermediary organizations such as local governments, business associations, and industry councils also playing a critical role. The intermediary agencies, such as the one at Suzhou Industrial Park established in 1994, provided institutional solutions to incentivize, regulate, and sustain successful partnerships.

(v) *Industry partnership.* To ensure the relevance of skills development, what policies were applied to promote employer-led training? In particular, amid rapid technological and market changes, what actual role did employers play? What incentives and regulations were applied? Under what conditions could such incentives and regulations work effectively?

The Korean government introduced compulsory vocational training legislation on relatively large employers in the 1970s, when the country transitioned to heavy and chemical industries from light industries. Despite resistance from the private sector, these policy measures were enforced while considering the unique political and economic environment in the country's early developing stages. In the PRC, in addition to promoting PPP with intermediary agencies, industry partnerships between dominant firms in strategic heavy industries such as railroads and the aeronautical industry made large investments in TVET institutions. The national TVET policy frameworks prepared by the government since the 1990s underpinned different forms of local industry partnerships to create synthetic knowledge.

(vi) *Comprehensive qualifications framework.* Once skills are developed, qualifications will be important in ensuring their social currency. How could a qualifications system contribute to economic development despite the absence of a comprehensive qualifications framework for all education and training programs? Under what conditions was such a contribution possible?

Different vocational qualifications systems were developed in both countries but these did not result in a comprehensive national qualifications framework. In the early stages of economic development, employers were looking for unskilled low-cost labor and did not necessarily look for vocational qualifications for employment. However, the absence of a qualifications framework did allow for the advantage of flexibility in creating new vocational qualifications to meet emerging demands. As both countries made industrial progress, the need for a comprehensive qualifications framework increased to manage confusion and inconsistencies under fragmented qualifications systems. A comprehensive qualifications framework, however, requires resources with high technical and institutional capacity.

(vii) *Pathways from general education to TVET. The development of TVET is deeply related to the development of the whole education and training system. What was the relationship between the development of TVET and general education? What effect did the development of general education have on the development of TVET?*

In the early stages of economic development, building a strong foundation in primary education helped poor but talented students to learn at TVET institutions. The rapid expansion of higher education during economic transformation, however, diminished the attractiveness of TVET in both countries, as more students took up the general education track to enter into higher education. In the PRC, there are attempts to make horizontal and vertical integration in TVET at the secondary school age level. However, because of the large differences in TVET and secondary schools, this integration is often challenging. The integration eventually takes place at higher education where there is enough space for enrolling students from both general education and vocational education tracks.

As discussed in Chapter 4, there is no "one-size-fits-all" solution, and the East Asian experiences show alternative approaches in developing TVET policies. These alternative approaches are made possible in the context of strong leadership and bureaucracy under developmental states, national security, cultural philosophy, and other historical factors unique to the PRC and the Republic of Korea. As a result, these TVET policies in East Asia do not necessarily work in other developing country contexts; thus, each developing country should come up with its own TVET policy measures. The building-block approach in TVET may still be a good reference but it does not serve as a necessary nor sufficient condition for successful TVET in other country settings.

Chapter 1
MODEL OF TECHNICAL AND VOCATIONAL EDUCATION AND TRAINING IN THE WORLD

The purpose of this study is to provide alternative technical and vocational education and training (TVET) models, building upon the experiences of the People's Republic of China (PRC) and the Republic of Korea. Before these two country TVET experiences are discussed, this chapter focuses on trends in the development of TVET worldwide, particularly the emergence of the building-block approach. It also analyzes TVET projects of the Asian Development Bank (ADB) from 2010 to 2020 to understand how the building-block approaches are used in projects. Typologies of TVET systems in the developed world are described using Germany and the United Kingdom (UK) to illustrate the diverse paths in developing a TVET system from a similar historical context. The chapter also examines failures in transplanting TVET models due to different historical and cultural contexts. Policy implications are then drawn as a springboard for the case studies on the Republic of Korea in Chapter 2 and the PRC in Chapter 3.

As a springboard for the East Asian case studies, this chapter poses the following seven questions:

(i) On the role of sector skills councils in identifying the labor market's skills needs, how was demand measured when the industrial base and employer roles were very weak and sector skills councils could not be formed?

(ii) When the industrial base was weak, how did the government pursue development of industry through aggressive skills development policies? In such cases, can TVET policy be said to take the leading role in future-oriented skills development, rather than a passive response to labor demand?

(iii) What was the status of TVET policy in the economic development strategy under the strong leadership of the government? What were the conditions for the government to successfully integrate TVET policy into its economic development strategy?

(iv) For government's skills development policy to be implemented effectively, enormous resources are needed, and public–private partnerships (PPPs) are critical to securing and enforcing them. How did this cooperation really happen? Did the public and private sectors have partnerships on equal footing? What was government's role in driving such cooperation? What incentives and regulations were applied? Under what conditions could such incentives and regulations work effectively?

(v) To ensure the relevance of skills development, what policies were applied to promote employer-led training? In particular, amid rapid technological and market changes, what actual role did employers take? What incentives and regulations were applied? Under what conditions could such incentives and regulations work effectively?

(vi) Once skills are developed, qualifications will be important in ensuring their social currency. How could a qualifications system contribute to economic development despite the absence of a comprehensive qualifications framework for all education and training programs? Under what conditions was such a contribution possible?

(vii) Finally, the development of TVET is deeply related to the development of the whole education and training system. What was the relationship between the development of TVET and general education? What effect did the development of general education have on the development of TVET?

1.1. Education for All versus Technical and Vocational Education and Training: Shifts in Priorities

Technical and vocational education and training, a term used interchangeably with "skills development," has come full circle in the last 60 years. It was first touted as a solution to the lack of development in Africa in the 1960s by leading academics such as Thomas Balogh, who warned the developing world not to "imitate the educational institutions, which have won a high reputation in Europe" (Balogh 1964, p. 10). That is, he urged against a disregard for technical education and against the exaltation of and overinvestment in liberal or general education, arguing instead that vocational education should be prioritized and integrated in general education systems if Africa were to achieve its rural renascence and reach a higher standard of living. This school of thought, albeit not without its critics (Foster 1965), caught the attention of multilateral institutions eager to lift Africa out of poverty, and TVET was prioritized in the policies and programming of bilateral and multilateral donors (Psacharopoulos 1991, Watson 1994, Ngcwangu 2015).

The United Nations Educational, Scientific and Cultural Organization (UNESCO) made its first TVET recommendations in 1962 (UNESCO 1962), and in 1976 held a conference on technical education urging developing countries to diversify secondary school curriculums and to develop technical education in basic and higher education (Hollander and Mar 2009). By the 1980s, TVET accounted for 40% of multilateral assistance to education (World Bank 1991, UNESCO-UNEVOC 2008). In 1987, during UNESCO's first International Congress on Technical and Vocational Education in Berlin, member states agreed that UNESCO should support development of an international center for TVET research and development; by 1991, the International Centre for Technical and Vocational Education and Training (UNEVOC) was launched (UNESCO 1989).

However, the tide had already begun shifting in the 1980s when it became increasingly clear that previous TVET investment, particularly in Africa, had not generated promised returns. Quantitative and qualitative research suggested TVET graduates were neither more employable nor better paid.

More importantly, the idea of producing labor for the economy through investment in TVET did not improve living standards and generate economic growth for society. Notably, the investment failures of Africa did not apply to all regions globally; in the industrializing nations of East Asia, TVET was an important factor in consistently low unemployment figures and was generating high returns (Tilak 1988, 2001). Mixed results notwithstanding, it appears that failures spoke more strongly than successes, leading some researchers to conclude that simply transplanting the model of TVET from developed countries to developing countries would not solve the problem. This was true even assuming that many fundamental foundational requirements for policy transfer, such as economic stability, were met. Context-specific solutions were needed (Okwuanaso 1984; Watson 1994; Carbonnier, Carton, and King 2014). As we demonstrate later, it has proved a hard-won lesson easily forgotten.

Other critics concluded that the training of skills should better be left with the market and private providers (Foster 1965, Psacharopoulos and Woodhall 1985, Psacharopoulos 1991). This shift was supported globally with the enshrinement of universal primary education in Education for All initiatives in 1990, the Dakar World Education Forum in 2000, and in the Millennium Development Goals (Psacharopoulos 2006). Ten years after the establishment of UNEVOC, UNESCO hosted a second International Congress on Technical and Vocational Education in 1997 in Seoul. Its recommendations on TVET strategy were expected to become the core of a new UNESCO global strategy for TVET but were instead largely ignored, something King (2009) attributes to the aforementioned shift in focus back to general education, and the relative convenience of the clear and simple aims of Education for All relative to the complex and comprehensive TVET recommendations from Seoul.

The result of this shift in focus was that during the period between the World Conference on Education for All in Jomtien (1990) and the Dakar World Education Forum (2000), financing for TVET dwindled if not disappeared entirely (Psacharopoulos 2006, Ngcwangu 2015) (Table 1 reviews World Bank education policy). The focus and influx of funding into universal primary education that ensued has translated into some laudable success in the run-up to the Millennium Development Goal targets of 2015. Yet, just as TVET cannot be manipulated to produce an economic effect out of school, the universalization of basic education cannot be expected to lead to the eradication of poverty (King and Martin 2002). The 2007/08 economic financial crisis brought this point to the fore with young and educated people among the big losers in the labor market (Carbonnier, Carton, and King 2014). While discontent with universal primary education and renewed interest in TVET did occur earlier, with aid from bilateral donors to TVET increasing threefold between 2002 and 2009 (McGrath 2012), it was the global economic crisis that cemented this policy shift at the multilateral and global level.

The World Bank Education Strategy 2020 paper states that "[t]he challenge is to give these young people appropriate opportunities to consolidate their basic knowledge and competencies, and then equip them with technical or vocational skills that promote employment and entrepreneurship" (World Bank 2011, p. 26; Ngcwangu 2015). UNESCO published two major reports on TVET and summoned the Third International Congress on TVET in 2012.

Table 1: World Bank Education Policy Swings Over Time

Time	Policy Priorities
1945–1963	No policy
1963–1987	Human resources, VOCED-oriented
1987–1990	Internal debates, confused
1990–1997	Basic, general education oriented
1997–2005	No clear priorities
2011–present	Renewed interest in TVET

TVET = technical and vocational education and training, VOCED = Vocational Education and Training Research Database.
Sources: G. Psacharopoulos and M. Woodhall. 1985. *Education for Development: An Analysis of Investment Choices*. New York: Oxford University Press; authors.

For the Asian Development Bank (ADB), nearly 50% of its education loan portfolio now is provided for TVET reforms, after decades of focusing on general education support. With the expected increase in TVET projects, it is critical to look into how TVET can be better designed and implemented.[1]

1.2. Trends in Technical and Vocational Education and Training Development—the Emergence of the "Building-Block" Approach

Examination of UNESCO TVET recommendations in 1962, 1974, 2001, and 2015 provides insight into the multilateral consensus on TVET as it has evolved. The original recommendations in 1962 were relatively rudimentary and often featured descriptions that now seem comparatively opaque. International comparison and relating practices to those of leading countries appears to be quite important within these recommendations and thus may have been regarded as a fundamental aspect of developmental programming in TVET. The 1974 recommendations expanded on those of 1962 considerably, further emphasizing lifelong learning, relevant economic forecasting, and systematic national research in TVET policy. Recommendations in 2001 do not appear to differ significantly from those in 1974, perhaps a result of that shift in focus toward general education.

The most notable shift in tone within the UNESCO recommendations occurs in 2015, where the previous structure of 1974 and 2001 are abandoned and more consideration is given to alternative TVET pathways and promoting TVET esteem in public discourse. Evidence of the renewed attention can be seen in the emphasis on TVET pathways and the relative lack of reference to "vocationalization" of general education compared to 2001. Interestingly, this document makes recommendations considering policies "according to their specific conditions, governing structures and provisions," suggesting support for more idiographic national approaches. Further, recommendations for international exchange seem somewhat de-emphasized in this document, and prior references to "international norms" were removed entirely. However, the degree to which these recommendations have been put into practice within the remit of UNESCO's own practice remains in question. For example, in 2018 UNESCO formulated 10 essential building blocks in TVET,

---

[1] The discussion about the ADB education sector program has been informed by internal discussions. For more information, readers can also consult the following documents: ADB (2011a, 2011b, 2014a, 2014b, 2016a, 2016b, 2016c, 2017, 2018a, 2018b, 2018c).

as shown in Table 2 (Todd and Dunbar 2018). Though elements of the 2015 recommendations are included (such as esteem for TVET), a universal, prescribed "building-block" approach seems contrary to the notion of affording nations the ability to formulate TVET systems according to the specific needs of their own environments.

Table 2: Building-Block Approaches for Technical and Vocational Education and Training

10 Essential Building Blocks in TVET	TVET Recommendations
1. Leadership and clarity of purpose across policy domains	1. National qualifications framework
2. Labor market relevance and demand-driven provision	2. Curriculum blending and ladders
3. Well-functioning partnerships and networks promoting access and equity, with partners adequately representing constituents' interests	3. Apprenticeships, internships, and on-the-job learning
4. High-performing, quality training institutions	4. Lifelong learning and adult and continuous education
5. Standardized quality assurance mechanisms and portability of qualifications	5. Partnerships with industry and the private sector
6. Stable and sustained financing	6. Mix of financing of TVET and equity
7. Well-functioning institutions, incentives, and accountability mechanisms	7. Linking TVET institutions with higher education institutions
8. Public esteem, strong graduation, and employment rates	
9. Availability of accurate data and information including labor management information systems	
10. Culture of policy learning and continuous improvement	

TVET = technical and vocational education and training.
Note: The 10 essential building blocks in TVET are from UNESCO and ILO (2018). The TVET recommendations are from Fawcett, el Sawi, and Allison (2014), with minor revision. Components 7 and 8 both relate to higher education and are therefore merged.
Sources: UNESCO and ILO. 2018. *Taking a Whole of Government Approach to Skills Development*. Geneva: United Nations Educational, Scientific and Cultural Organization and International Labour Organization; C. Fawcett, G. El Sawi, and C. Allison. 2014. *TVET Models, Structures, and Policy Reform: Evidence from the Europe and Eurasia Region*. Washington, DC: United States Agency for International Development.

Notably, Balogh (1964) cited unemployment as the reason to prioritize TVET over general education. Five decades later, after we have tried both approaches, (i.e., TVET and Education for All), we are back to square one. The real problem, however, is that we have forgotten the lessons along the way and hope that the new international norms or best practices in TVET, largely based on the experiences and practices of developed countries, can once more be the silver bullet for developing countries. This trend is not simply contained within multilateral organizations; for example, consultants for the United States Agency for International Development summarized eight key components of effective TVET systems (Table 3), drawing primarily on the models of France, Germany, and the United Kingdom.[2] McGrath (2012) states that these approaches are similar because they all draw heavily on the new

---

[2] As noted, the ADB education portfolio is based on internal discussions. See footnote 1 for ADB education projects.

public management paradigm, an approach to public administration that emerged in the 1980s and aimed to channel techniques from modern science and the private sector to increase the efficiency of public agencies (Gow and Dufour 2000). He simplified them into five principal tools: (i) systemic (and sometimes sectoral) governance reforms, (ii) qualifications frameworks, (iii) quality assurance systems, (iv) new funding mechanisms, and (v) managed autonomy for public providers. These tools offer several theoretical advantages, including making the system more flexible for TVET students and making funding more outcome-oriented. However, perhaps the broadest focus of this paradigm is to transfer more agency away from public agents to stakeholders, allowing for a more demand-responsive and autonomous system.

Table 3: Key Areas in Technical and Vocational Education and Training Transfer

Institutional Mismatch	Cultural Mismatch
Relevant to the way important national institutions and structures of governance are organized.	Relevant to the distinct place of TVET within a nation's culture and history.
Power Differentials	**Sociological Mismatch**
Relevant to the imbalance in decision-making power between developing nations where partnerships may exist with multiple donors using varied approaches.	Relevant to the particular social structure of a country and how it can interfere with a country's ability to borrow from other systems.

TVET = technical and vocational education and training.
Source: T. Lewis. 2007. The Problem of Cultural Fit—What Can We Learn from Borrowing the German Dual System? *Compare: A Journal of Comparative and International Education.* 37 (4). pp. 463–477.

This seeming consensus is rather strange as it not only neglects the earlier lessons on the importance of context in education policy recommendations, but it is also perplexing against the background of diverse models of skills formation systems in the developed world. The approach in multilateral organizations to developing TVET systems has followed a certain pattern over the past few decades. A select group of features inspired by the success of certain models from developed countries manifest often in TVET projects. For example, out of the 29 ADB loan projects (2010–2020) analyzed within this research, 19 featured at least one of five features: (i) national vocational qualifications framework, (ii) public–private partnerships (PPPs), (iii) competency-based curriculums, (iv) labor management information systems, and (v) sector skills councils. The most prominent are discussed below.

National Vocational Qualifications Frameworks

National vocational qualifications frameworks are used to certify the quality and training content of vocational training provided by all recognized training organizations. These frameworks assure those who undertake vocational training that their qualification will be recognized by potential employers, and assure employers that a prospective employee has the skills they require (Williams and Raggat 1998). In many developing countries, numerous actors perform TVET; thus, the quality and degree of training has varied greatly, leaving TVET trainees unable to adequately verify their training to potential employers. This failure often drives more students toward a university education, where their credentials can at least be recognized. National vocational qualifications frameworks are typically developed in consultation with relevant private sector actors to ensure that training meets industry demands. In this sense, the frameworks are typically developed in accordance with efforts to generally improve the quality and relevance of training, as in ADB projects executed in Indonesia (2018),[3] and more. Reasons cited for the inclusion of national vocational qualifications frameworks included a high dropout percentage among TVET students or frequent inability for TVET graduates to find employment using their qualifications. In Cambodia (2014),[4] one such framework was devised to expand the links between competency qualifications and additional educational pathways to promote continual skills development and lifelong learning. In Mongolia (2014),[5] another framework was developed given low labor force participation among TVET graduates, secondary schools, and universities that created a credit transfer mechanism to link all these types of educational pathways.

Competency-Based Curriculums

In competency-based training, the curriculum is built around learning outcomes (knowledge, skills, attitudes, etc.) as opposed to specifying the content of what will be taught in the learning environment (UNESCO International Bureau of Education n.d.).[6] The greater emphasis on learning outcomes and reduced emphasis on the teaching method allows teachers the flexibility to adjust teaching to better suit the learner. Competency-based training is thus a natural fit within TVET, as students are being prepared for specific roles that will require proficiency in certain tasks. It also seems like a logical fit given the potential profile of TVET students, who often find themselves enrolled in TVET schools due to a history of poor academic performance, and where teachers may need to use a variety of teaching techniques depending on student's needs. The logic for including competency-based training is relatively self-explanatory, as it links competencies to qualifications, which thus assures employers of their expectations for employees. Thus, such training is quite often developed alongside national vocational qualifications frameworks, and is typically linked to a competency-based curriculum. The rationale for including competency-based training in some projects typically stems from issues related to quality and relevance of training, as with the frameworks. Specifically, competency-based training may be used to remedy TVET systems where training is not deemed practical enough and focuses too much on theory, as was the case in developing countries. Competencies are developed around specific occupations and demand the mastery of a particular set of competencies, thus qualifying a person to perform a certain job. Occupation profiles may be built around certain occupations deemed to have suffered from poor TVET governance and low skill-relevance, as was the case of Tajikistan (2018).[7]

---

[3] ADB. Indonesia: Advanced Knowledge and Skills for Sustainable Growth Project. https://www.adb.org/projects/50395-006/main.
[4] ADB. Cambodia: Technical and Vocational Education and Training Sector Development Program (TVETSDP). https://www.adb.org/projects/46064-002/main.
[5] ADB. Mongolia: Skills for Employment Project. https://www.adb.org/projects/45010-002/main.
[6] International Bureau of Education. Competency-Based Curriculum. http://www.ibe.unesco.org/en/glossary-curriculum-terminology/c/competency-based-curriculum.
[7] ADB. Tajikistan: Skills and Employability Enhancement Project. https://www.adb.org/projects/51011-001/main.

Public–Private Partnerships

Public–private partnerships, meanwhile, are partnerships between an agency of the public sector and the government in the delivery of goods or services to the public (Britannica 2019). The different types of PPPs can serve a variety of functions. While many PPPs are purely financial, where the role of the private agent does not go beyond the provision of funds—many others are designed to leverage existing capacity of certain private agents.

As per the ADB *Public–Private Partnership Handbook*, the "partnership should be designed to allocate risks to the partners who are best able to manage those risks and thus minimize costs while improving performance" (ADB 2008, p. 1). In this, private agents may broadly be involved in some administrative or managerial capacity, depending on the needs of the project and their ability to contribute relative to that of the public sector. These partnerships are aimed at cultivating a mutually beneficial relationship between public and private actors, and play an increasingly important role in promoting decent work around the world (ILO 2021).

Public–private partnerships have recently become a point of stronger emphasis in multilateral portfolios as they may be able to narrow the gap between traditional donor-based funding and projected future costs of infrastructure. The most common reason cited in this research for inclusion of PPPs is their ability to increase industry involvement in the execution and design of training, theoretically making it more relevant to the needs of industry and thus making TVET graduates more employable. For example, PPPs were included in the 2020 TVET development project in Georgia where "extent of staff training," "quality of VET,"[8] "skill sets of graduates," and "ease of finding skilled employees" were cited as areas of severe disadvantage.[9] A survey of Georgian employers indicated they generally believed that the educational system was not responding to their needs. PPPs also help TVET institutes gain access to equipment and facilities to effectively train students in certain areas. For example, in a 2018 TVET project executed by ADB in Shanxi,[10] PPPs allow potential private actors to help "design, build, manage, and operate practical training facilities" in key domestic sectors such as artificial intelligence, big data, e-commerce, and so on. In 2011, projects in Nepal[11] and in 2016 projects in Madhya Pradesh, India,[12] PPPs were cited as important for service design and delivery in training provision and industry-aligned skills certification. Clearly, PPPs are not merely used as an additional stream for funding TVET projects, but are an important mechanism for allowing private sector input into TVET systems.

Labor Management Information Systems

Labor management information systems allow education and labor policies to better reflect labor market needs, mainly to prevent national skills mismatches (ILO 2016). These systems include

---

[8] The terms TVET and VET are used interchangeably.
[9] For the 2020 TVET development project in Georgia, see ADB. Georgia: Modern Skills for Better Jobs Sector Development Program, Subprogram 1. https://www.adb.org/projects/52339-001/main.
[10] For labor management information systems used in the PRC, see ADB. China, People's Republic of: Shanxi Technical and Vocational Education and Training Development Demonstration Project. https://www.adb.org/projects/51382-002/main.
[11] For Nepal, see ADB. Nepal: Skills Development Project. https://www.adb.org/projects/38176-012/main.
[12] For Madhya Pradesh, India, see ADB. India: Madhya Pradesh Skills Development Project. https://www.adb.org/projects/48493-001/main.

everything from acquiring, storing, and analyzing data, as well as using data to make informed decisions about the labor market. Labor management information systems are important for monitoring trends in employment levels in a given region and analyzing the sectors and skills most in demand. Within ADB projects, such systems are commonly associated with aligning the skills created within the TVET system with the demands of industry. In 2014 in the PRC, labor management information systems were used in Chongqing; for example, where a need to standardize training across regions was met with development of the systems to allow better interregional data-sharing.[13] Furthermore, the Chongqing government used a labor management information system to integrate the vocational training system into the larger social security framework, which could help enrollment. Other projects include one in Timor-Leste in 2011, where data was only collected on the supply side from TVET training providers, requiring upgrades to include the demand side from employers. Labor management information systems are a common feature of projects to increase demand-driven TVET training.[14]

Sector Skills Councils

Sector skills councils or "industry skills councils" are used to include the participation of particular industries within the mechanisms of a TVET system. Their value is in "specifying the nature of the skills that an industry sector needs" from the perspective of those directly "exposed to the competitive pressure of domestic and international markets" (ILO 2019, p. 2). As noted, it is often a priority that industry be given a direct voice within the design and execution of a TVET system alongside the likes of data monitoring, to fully ensure training is meeting the needs of industry. Sector skills councils can be useful for capturing the nuances of the needs within particular sectors. For example, a 2016 project in India that focused on inclusive growth used such councils to devise recommendations for gender equality in the workplace in the health care and construction sectors. Sector skills councils have also been used in projects to establish or improve their role, including linking the central TVET authority to suitable demand-side employers, to be able to collect relevant data. A project implemented in India in 2017 and another in Cambodia in 2014 also mention the use of sector skills councils in identifying and providing internship and placement opportunities for TVET students.[15]

Such features have been inspired by strong organizations and well-established PPPs upholding successful TVET systems around the world; therefore, the intentions behind their inclusion are logical. For example, the idea of competency-based training can be compared to the workplace-led competency approach in Germany, where hundreds of listed occupations are defined by their competency profile. Public–private partnerships also seem to represent an effort to mimic how Germany funds TVET training, where companies voluntarily supply training given the protections granted by public institutions (Deissinger 2015). The UK's National Vocational Framework also seems to be the clear inspiration behind the national vocational qualifications frameworks included frequently in multilateral projects (Young 2011). However, the complex reality of these

---

[13] For the People's Republic of China, see ADB. Chonqing Innovation and Human Capital Development Project https://www.adb.org/projects/50222-002/main.

[14] For labor management information systems used in Timor-Leste, see ADB. Timor-Leste: Mid-Level Skills Training Project. https://www.adb.org/projects/45139-001/main.

[15] For Himachal, India, see ADB. India: Himachal Pradesh Skills Development Project. https://www.adb.org/projects/49108-002/main.

features' origins must be considered. Technical and vocational education and training systems, and education systems more generally, are the product of centuries of cultural, sociological, economic, and political factors that have shaped the system and the elements that support its existence. Attempts to instigate the development of similar features in developing countries may seem logical, but true replication is impossible, suggesting a fundamental flaw in this approach.

1.3. Typologies of Technical and Vocational Education and Training Systems in the Developed World

Alongside theoretical research on TVET, many attempts have been made to create broad descriptive "typologies" of TVET that could help to conceptualize different types of systems, including Sung, Turbin, and Ashton (2000). Busemeyer and Trampusch (2012a) create an influential typology that examines TVET systems on two dimensions: firm involvement in (initial) TVET, and public commitment of the state to the provision and financing of TVET. The first dimension concerns the extent to which employers invest in training, resulting in the provision of nonfirm-specific technical skills, overcoming coordination risks (described later). The second dimension relates to the extent of state support for vocational education as a credible alternative to mainstream general education. When examined through these dimensions, four possible TVET systems emerge: segmentalist, collective, liberal, and statist skill formation regimes (Figure 1).

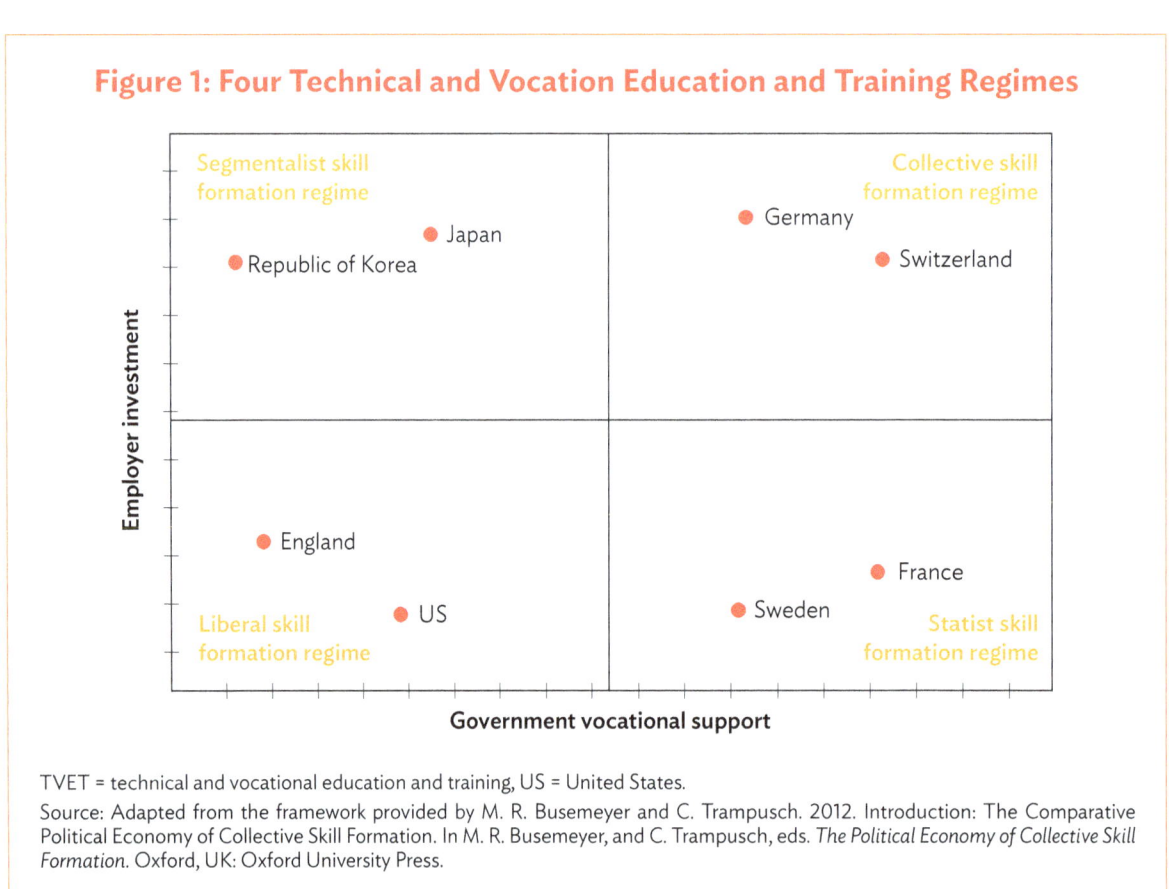

Figure 1: Four Technical and Vocation Education and Training Regimes

TVET = technical and vocational education and training, US = United States.

Source: Adapted from the framework provided by M. R. Busemeyer and C. Trampusch. 2012. Introduction: The Comparative Political Economy of Collective Skill Formation. In M. R. Busemeyer, and C. Trampusch, eds. *The Political Economy of Collective Skill Formation*. Oxford, UK: Oxford University Press.

(i) *Segmentalist skill formation regime.* This features high employer involvement in vocational training but a lack of state promotion and standardization of vocational training. This undermines public attraction to vocational education and hinders how transferable training is, and it is not firm-specific. This type of system can be seen in Japan and previously in the Republic of Korea.

(ii) *Collective skill formation regime.* This is defined by high employer involvement as well as high commitment to promoting vocational education from the state. State commitment to vocational education is typically maintained through a corporatist governance framework, which involves everything from standards creation, practice monitoring, and collaborating with employer organizations to keep up with industrial change. The PPP provides assurances that allow employers to participate in vocational training, thus employers and the labor force can mutually benefit from the system. The German "dual system" is the quintessential example of such a regime, where students may apprentice for one of over 300 registered occupations in a private enterprise for 2–3 days a week, spending their remaining days learning in a vocational school and eventually graduating as a licensed professional (Deissinger 1994, 2002b).

(iii) *Liberal skill formation regime.* This is defined by low investment in vocational training by employers and little state effort to promote vocational education as an alternative. From a human capital perspective (Becker 1964, 1993), employers are unlikely to invest in general skills provision due to the potential for trained employees to be poached, therefore "training" is often on-the-job and focused on firm-specific skills. Workers are less likely to invest in their own skills, as they have to pay for this themselves through dubious training providers. The UK is a modern example of such a regime, where the government has not established an effective state-provided TVET pathway and a big variety of relatively unregulated training providers dominate the TVET landscape (Young 2011).

(iv) *Statist skill formation regime.* This is defined by a high commitment to promoting vocational education, where suitable institutions and standards are built accordingly but employers' investment in vocational training is low. High state involvement can mean that keeping up with industrial trends is difficult; therefore, training struggles to meet the demand of the current labor market and encourages the participation of private enterprises. From a human capital perspective, a dominant state-sponsored TVET pathway that does not produce good results can drive more people to attend university due to perceived higher reliability.

Some research on TVET systems and development has focused on their origins and attempts to explain their growth. One comprehensive study (Danforth 2014) examines the explanatory power of three different theoretical perspectives. Power resources theory examines the development of institutions as a result of class dynamics between upper and lower classes; for example, the initial reduction in power of the lower classes stemming from the industrial revolution and the devaluation of skilled labor. The second theoretical perspective, varieties of capitalism, examines the role of social policy in reducing "coordination risks" between private and public actors within the TVET system. For example, the success of reducing coordination risks for private employers, like employee poaching, has been a major instigator of success within the German dual system. These perspectives offer certain value in explaining the evolution of TVET systems. However, certain factors are not covered within their framework.

The final perspective included by Danforth, historical institutionalism, seems to offer the strongest explanatory power in TVET system origins. This perspective focuses on the unique cultural and historical circumstances of particular regions in the development of education systems, a perspective that has been supported elsewhere in literature (Deissinger 2015, Wilson 2001). This perspective emphasizes timing, whereby culture and other factors can become extremely influential through historical windows of volatility. This notion seems to fit TVET development well. For example, one can see how the response to the Industrial Revolution in Germany and the UK was greatly influenced by their respective cultures, for very different developmental trajectories.

1.4. Country Cases—Role of Culture in Technical and Vocational Education and Training Within Germany and the United Kingdom

While typologies may help understand and categorize the TVET systems of the world, it neglects the origins of such systems, meaning that the circumstances underpinning TVET development are underdiscussed. As noted, cultural context plays a strong role in how much states can mimic other TVET systems. It is not enough to merely be able to describe TVET systems; it is also imperative to know how these regimes came to be and the dynamics sustaining them. Focusing on the concept of "historical institutionalism," this section outlines the diverging histories of TVET in Germany and the UK, starting from their relatively common beginnings during the "guild era," up to the stark contrasts in their modern approaches.

Phase 1: The Era of Guilds and Apprenticeships

The roots of both German and British vocational training (Figure 2), as is generally the case across Europe, can be traced back to guild systems beginning around the 12th century (Deissinger 1994, 2002a). These guilds controlled the skilled labor force through their system of "apprenticeship," whereby any person who wanted to be a craftsperson would train under the supervision of a "master" for roughly 7 years to gain "freedom of the trade." Guilds were an extremely important aspect of medieval society, as members often formed a tight-knit community. Ruling governments saw guilds as a threat to their power as they exerted significant control over political and economic affairs, through powerful lobbying and the ability to influence labor markets (Wollschläger and Guggenheim 2004). The story of the guilds in Britain is almost identical to Germany's, though power struggles between British guilds and the British Crown may have been more severe (Hoogenboom et al. 2018). Thus, generally it is fair to say that the mechanism of skills provision in both countries was relatively similar at that time.

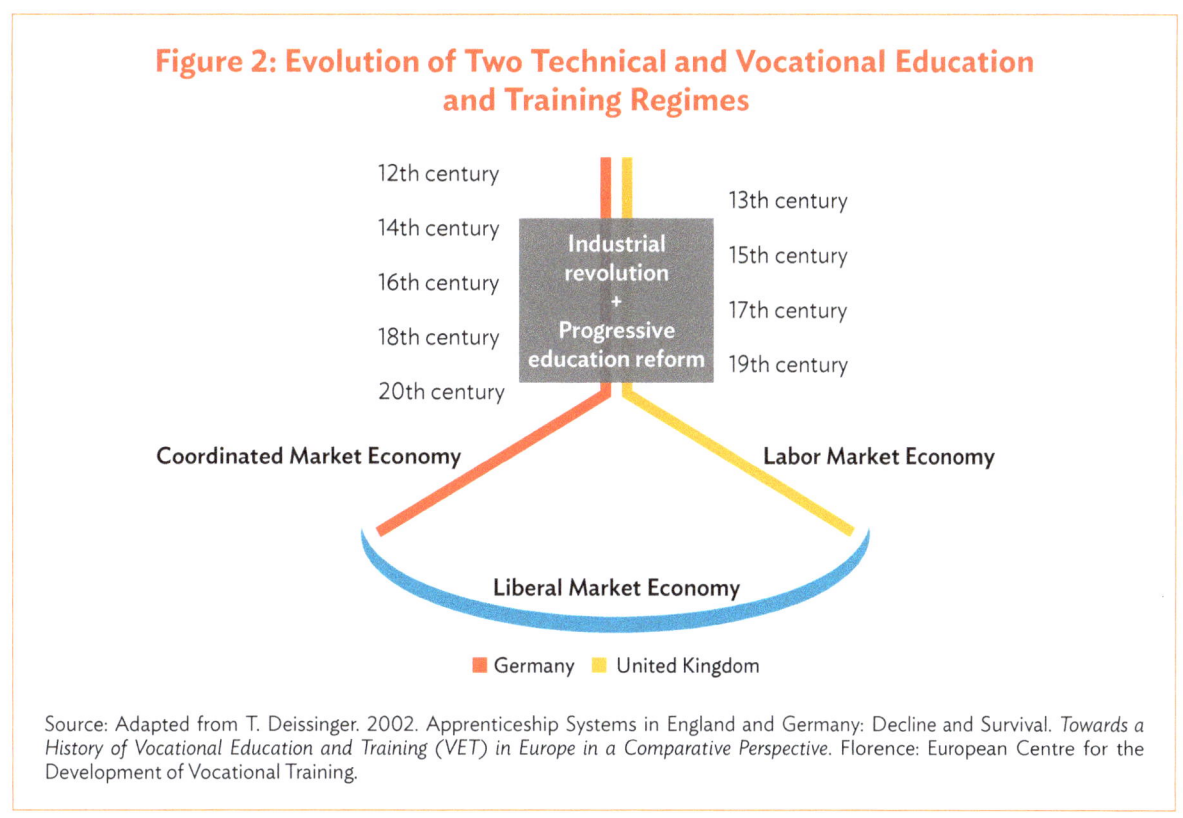

Figure 2: Evolution of Two Technical and Vocational Education and Training Regimes

Source: Adapted from T. Deissinger. 2002. Apprenticeship Systems in England and Germany: Decline and Survival. *Towards a History of Vocational Education and Training (VET) in Europe in a Comparative Perspective*. Florence: European Centre for the Development of Vocational Training.

Phase 2: Industrial Revolution and Collapse of the Guilds

After the Industrial Revolution and the introduction of complex machinery, the need for vast numbers of low-skilled labor increased and for high-skilled labor decreased. Guild influence over highly skilled labor was thus much less important, and governments used this opportunity to reduce their rights and expand freedom of entrepreneurship to more people (Kieseweiter 1989). Several acts in the mid-19th century granted freedom of trade to "free masters" who had no relationship with the guilds and removed the guilds' status as public institutions. This represents the lowest point for the guilds within the context of German history. However, a huge working class now existed, united through worker organizations and simple class dynamics, fomenting socialist upheaval. The government of Wilhelm II responded to this pressure by reintroducing protections for guilds and highly skilled craftspeople (Blackbourn 1977).

In comparison, the Industrial Revolution was born in the UK, and the extent to which economic liberalism influenced policy was most severe here. After the guilds were abolished in 1814, use of full-time child labor increased massively and these children could now earn as much as their parents in factories. Training was poor to nonexistent and only carried out on a job-specific basis, and a "labor aristocracy" served to restrict access to well-paid positions and keep low-level workers in place (Deissinger 1994, 2002a). Without the apprenticeships, private schools were left as the only route through which low-class British citizens could invest in their human capital. Unfortunately, private schools were often run by religious groups focused only on general skills like reading and writing. There was also an existing general aversion to skill-based education within British culture

(Abbot 1933). With the rise of progressive liberalism, several acts would be passed to increase school attendance (the Fisher Act, 1918 and the Butler Act, 1944) but would ultimately fail due to the still-pervasive liberal attitudes in British policy making (Perry 1976). This shows how unique cultural factors within these two countries caused diverse responses to the same phenomenon and launched two distinct development trajectories in education and skills provision.

Phase 3: Postindustrialization Expansion of Education

Socialist influences on German social policy would eventually lead to attempts to create an expansive education system. George Kerchensteiner was a very influential educationalist in Germany, who saw education as a tool to imbue students with a set of moral and civil values that could create loyal workers and perhaps dissuade citizens from mass protest (Winch 2006). Thus, the very roots of the German TVET system are connected to a concept of "vocation" that goes beyond the classroom and frames vocational learning within the broader notion of national pride and civic duty. Schooling became mandatory in 1938, and the vocational "continuation schools" became much more widely attended (Abel and Groothoff 1959). Eventually, trades were systematically classified, apprenticeships were standardized, and formal examinations were introduced. Private enterprises were encouraged to participate by strong state protections and a strong voice for trade unions in training design (Blankertz 1982). This movement within German education would gain momentum as other 20th century political regimes capitalized on the "social being" aspect to promote stability and national identity, and the "dual system" was legally ratified through the Vocational Training Act, 1969 (Kümmel 1980, BiBB 2014). Today, the dual system still enjoys strong attendance figures.

In the UK, after the liberalist failings of creating a strong education system, a huge slump of in-company apprenticeships in the 1950s led to the government's decision in 1964 to finally abandon *laissez-faire* in its skills approach. The efforts of the British government to establish an effective vocational education system beyond this point are marked by the Industrial Training Acts in 1964 and the Employment and Training Act in 1973, and further reforms introduced in the era of Prime Minister Margaret Thatcher. The "Modern Apprenticeship Scheme" was cofinanced by the public sector, which was linked to a national vocational qualifications framework system established in 1986. However, while efforts have increased in the past few decades on behalf of the British government, fundamentally, resistance remains to becoming integrally involved in the actual execution and regulation of training and truly making the TVET system built on PPP (Deissinger 1994, 2002a). For example, companies still do not have to fix a definite training duration and the regulations around training licenses are still relatively soft (Young 2011).

It is clear that many unique historical and cultural factors in these two nations have significantly influenced the development of each education system, including TVET, and it seems that historical institutionalism best explains their characteristics. Referring back to the typologies of Busemeyer and Trampusch (2012a), the UK may now be classified as a "liberal skills formation regime" while Germany may now be classified as a "collective skills formation regime," existing on the exact opposite ends of the spectrum. This major gulf has emerged in spite of the homogenous starting points both countries shared during the guild era in the 12th–16th centuries. This demonstrates not only the diversity of TVET systems, but the degree to which indirect factors can result in such severe divergence. Historical institutionalism as highlighted by Danforth allows us to understand that even within the relatively confined area of Western Europe, no two countries are the same. This is evidenced by the continual influence of different cultural approaches to social policy, decision-making by rulers, and other such unpredictable and unmeasurable qualities. This perspective may not offer neat, conceptual classifications, but it seems that the specific historical legacy of each TVET system and cumulative effect of each twist and turn therein is too complex to simplify in such terms.

1.5. Limited Successes in Transplanting Models of Technical and Vocational Education and Training

In examining the German and British cases, specific areas of divergence are evident that may need to be considered when examining the potential fit of one type of dual system within a new cultural and historical context. Lewis (2007) identifies four key areas for consideration (Table 3).

Examples of these "mismatches" can be found in practice. For example, in the late 20th century, the British government attempted to institute features that resembled those of the German dual system, but due to the "cultural mismatch," whereby British private enterprises lacked the "training culture" to effectively adhere to the standards laid out by planners, the British government would eventually pursue a TVET system that diverged strongly from the dual system. Policy makers in the United States (US) considered implementing a dual system, but this sparked a debate about the suitability of such a system for American demography. Some felt creating such a pathway could deepen pre-existing racial inequalities, making the US an example of a "sociological mismatch" for the dual system (Deissinger 2015, Kliebard 1999, Oakes 1985).

One can imagine, compared to the relatively stable and homogenous backdrops of western economies, how attempts to replicate the dual system in developing countries, with their limited public and private infrastructure, might be even more difficult. Indeed, Wilson (2001) highlights the use of dual systems as something of a panacea to cure the ills of skills development globally. In Botswana for example, Wilson, Kennedy, and De Jocas (1999) cited "cultural determinants" such as the high status of the *meister* (master) affecting the flexibility of the educational institutions in adapting their offering to industry's needs as leading to the eventual failure of an attempted dual system. Mittman (1998) described a dual system-inspired pilot project in Costa Rica that linked enterprises to the national vocational training system, which proved unable to provide general skills training that was not firm-specific. In India, one state introduced a system to emphasize practical aspects over theory but was seen to have failed due to insufficient equipment and facilities, unqualified teachers and instructors, an inadequate training system for teachers, a lack of linkage with industries, the inferior status of vocational education, and the ineffectiveness of centrally sponsored schemes (Wilson 2001). These are just some examples of the difficulties encountered in dual system transfer within the developing context.

1.6. Policy Implications and Conclusion

On Borrowing

The implications of our findings are not that wholesale borrowing should be ceased as a practice entirely, but that attempting to completely replicate different systems in new regional contexts should be replaced with a more nuanced approach. Singapore is one country that has demonstrated positive results based on modeling inspired from the German TVET system (Seng 2011). Lewis (2007) provides four examples of how borrowing could be better used in the context of international development:

(i) *Partial borrowing.* This can be used to identify aspects of another TVET system that may seem to fit best with a nation's own particular needs.

(ii) *Pilot borrowing.* This can be used to examine how viable system changes that are either drastic or moderate work on a small scale before full-scale implementation.

(iii) *Customization.* This involves adopting a system but tailoring it to suit the needs of the local context, which may be particularly useful to developing countries whose TVET structure is highly underdeveloped and in need of drastic changes.

(iv) *Conceptual borrowing.* This uses more abstract modeling, whereby the intentions or essential conceptual dynamics of the system may be used as inspiration for a TVET system.

Whether developing nations choose to borrow from other models or decide to use a more organic approach based on their own individual needs, it is important that attempts to transplant features of foreign models directly into new contexts are eliminated from educational development.

On Importance of History and Culture

As noted often, perhaps the most important implication in the origin and continued successful development of TVET systems is the importance of cultural and historical factors. Certainly, there are features of a TVET system that broadly make sense regardless of context. For example, it seems clear that an emphasis on private sector involvement within the continual design and execution of training is an important ingredient for success. However, instituting such a dynamic is easier said than done, as is evidenced by the many examples of countries that appear to have attempted and failed to do this. The real challenge in developing an effective TVET system is finding a solution that incentivizes participation on the part of the government, the private sector, and the public. To do this, solutions must be crafted around the interdynamics of these three agents, which are unique to every country in the world. Therefore, simply borrowing an entire system that has been shaped around a completely different historical and cultural context will likely rarely have the desired results.

As Deissinger (2015) points out, attempts to transfer the dual system may eventually randomly succeed in practice, but they will never succeed in creating a dynamic identical to that of Germany. The problem is not with the dual system per se, but in the lack of consideration for the influence of national history and culture. For a government to identify a system that suits its needs, it must, to repeat Sadler (1912), look inside the "homes of the people," and attempt to discover the "intangible, impalpable, spiritual force" that governs the dynamics of their own country, and build from there.

The next chapters of this study will look at the example of two countries—the PRC and the Republic of Korea—that have developed their TVET systems with consideration for the nuanced complexities of their own environment and attempts to align skills development to broader national strategies. This intends to demonstrate how the two countries have defied conventional wisdom and come up with homegrown solutions to a set of common challenges in technical and vocational education and training.

Chapter 2
EXPERIENCE OF THE REPUBLIC OF KOREA

As discussed in Chapter 1, this case study on the Republic of Korea will delve into seven questions related to (i) sector skills council, (ii) skills development for the future, (iii) TVET policy in economic development, (iv) public–private partnership, (v) industry partnership, (vi) vocational qualifications framework, and (vii) pathways from general education to TVET. Answers to each of these questions will have three components: first, a background description of each policy or activity; second, a description of the specific development process or progress of each policy or activity; and third, a discussion reviewing the conditions for success. Through this three-dimensional critical review, it is hoped that the experiences of the Republic of Korea can be interpreted appropriately and useful implications drawn.

In general, vocational education and training is regarded as an important policy in achieving social integration and economic growth through the development of vocational skills for the socially underprivileged and enhancement of business competitiveness. Accordingly, developing countries, which are deeply interested in economic growth, show considerable interest in activating vocational education and training policies in their countries. However, it has not been easy for these countries to advocate such policies. First, vocational education and training is a learning program that requires a large amount of money for practical training to equip people with the competencies necessary for actual work. Second, although significant money could be spent, it would not be easy for such investments to reap the originally intended effects unless some conditions were met. For instance, if vocational education and training is carried out without knowing what skills are actually required in the labor market, huge investments can simply lead to huge losses. For these reasons, the vocational education and training policy, despite being highly important, is a policy that has always been a target of heated controversy in its actual implementation.

The Republic of Korea was no different. Before economic development in the 1960s, the country was one of the poorest in the world. It suffered from a serious civil war only 5 years after liberation from Japanese colonial power, which caused enormous human and material damage. Most industrial facilities were burned down and most businesses were extremely small. Moreover, economic growth based on natural resources was not an option due to its scarcity. However, the Republic of Korea has constantly emphasized the importance of learning for the past 1,000 years and learning aspiration has been established in the "social DNA" of Koreans. As such, the Korean government has adopted economic development policies that have actively utilized its excellent human resources since the 1960s, and vocational education and training policies have contributed significantly in this regard by fostering a competent industrial workforce as part of such policies.

However, development of vocational education and training has never been smooth. Funds for it have always been limited and constant controversy has existed over whether such investments are nurturing truly needed industrial staffing. In addition, in the formation of the institutions that are the focus of this book, there was neither the institutional structure needed for planning appropriate vocational training policies, nor the expertise to execute such policies.

To overcome this situation, the Republic of Korea benchmarked various vocational training policies from developed countries, alongside the mobilization of massive financial support for them. First, the country looked into Japan's vocational training system, which provided the basis for modern education and training—despite its colonial past—especially during the system's formative stage. For example, the Japanese vocational training act was referred to during enactment of the Vocational Training Act, 1967, the first law to build the basic framework for vocational training policy. The Japanese system was also considered for the development of vocational training standards, the design of the qualifications examination system, and the application of accreditation standards across private vocational training institutions. In this regard, the Japanese experience could be considered the most important impact on the Republic of Korea's vocational training system. However, the Korean government had to make significant modifications to reflect the large developmental gap between the two countries and took into consideration cases from countries other than Japan.

In particular, the Republic of Korea introduced a US-style education system after its liberation from Japanese colonial rule in 1945 and received support from various western countries such as Belgium, Germany, and the US in expanding public vocational training. In particular, German advisors provided very useful practical advice about the design and operation of vocational training facilities during relatively long (i.e., sometimes more than 5 years) in-country stays. However, their contributions seem to have had limited impact on the overall institutional framework of vocational training in the Republic of Korea due to the large difference in social and economic conditions between these countries. For example, while the Republic of Korea received substantial support from Germany, the main channel of training skilled workers had been maintained through schools or institutions rather than German-style dual apprenticeship, which required close partnership between schools or institutions and employers. Although the Korean government attempted to benchmark such dual apprenticeship training after the late 1990s, dual training was not considered a desirable model in the Republic of Korea, especially during the early stage of its economic development, when employers did not show much interest in securing skilled workers.

Therefore, the vocational training system of the Republic of Korea did not follow the development path of any particular country, but instead adopted approaches from various countries, sometimes with significant modification, according to its unique needs during each period. This can be viewed as an institutional learning process that inevitably involves trial and error, an experiment with huge cost consequences.

In the following subsections, the Republic of Korea's pursuit of vocational education and training policies through economic development will be explained from the viewpoint that each country should constantly search for the most appropriate methods for its own situation. In particular, while answering the questions raised in Chapter 1 about the conditions for success in vocational education and training, this chapter aims to inspire insights for developing countries in designing and implementing policies that fit their own circumstances. That is, as the Republic of Korea's experiences have been based on its unique socioeconomic context, it would be inappropriate for developing countries to directly imitate these experiences. Instead, by objectively looking at the Republic of Korea's successes and failures on specific policy issues, each country would be able to obtain useful implications for designing policies that fit their circumstances.

2.1. Identification of Labor Market Skill Needs When Sector Skills Councils Could Not Be Formed

Background

Before economic development began in earnest in the Republic of Korea, its industrial base was extremely vulnerable. For instance, when considering the distribution of companies by industry prior to economic development in the 1960s,[16] 6,842 businesses operated in 1959, of which 4,198 were in manufacturing (61.4%). Even among manufacturing companies, small and medium-sized enterprises (SMEs) were the overwhelming majority, even considering the high labor intensity in the manufacturing sector at that time. That is, only 29 manufacturing companies had 201 or more employees, and 162 companies had 151–200 employees. According to the Economic Planning Board (EPB), most companies (1,330), had just 51–100 employees and 2,099 just had 5–50 employees. In a survey of working conditions of the entire workforce (EPB 1961), among the 8.52 million employed people in 1960, employees with wage contracts were 0.98 million, only 11.6% of those employed. This premodern employment structure was dominant throughout the Korean labor market at the time.

It was thus impossible for government to identify the skills needs of industry through active engagement of employers in the skills system. First, only a limited number of employers were able to secure enough expertise in human resource development within companies. Second, most employers were uninterested in the acquisition of skills through long-term vocational education and training because production was only dependent on low-wage unskilled labor. As such, even if employers voiced their specific needs, they could not be regarded as appropriate demands on vocational education and training.

Accordingly, from the beginning of economic development, the Korean government set the contents of education and training by itself in consultation with private experts and professors instead of relying on a collective channel of employers such as sector skills councils. Of course, that process did not go as smoothly as the government had planned, and the results were by no means ideal. However, such attempts seemed inevitable given the lack of channels to properly identify the demand for skills in industries.

---

16   Using the Korea National Statistical Office Database.

The following examines the Korean government's attempts to identify and make concrete the demand for skills during the early stages of its industrialization, focusing on the set up and subsequent development of vocational training standards as supervised by the Labor Administration.

Progress

With the limited role of employers, government inevitably led the development of content for vocational education and training. For example, in preparation for the vocational training project that was implemented from 1967, the Labor Office designed vocational training courses and curriculum standards (Seo 2002) using data from the Korea Human Resources Research Report and the First Five-Year Plan for Technology Promotion prepared in 1961 by the EPB.[17] The Labor Department identified statistics on the number of workers by type, region, age, and education of technicians and skilled workers using these data, and assessed the extent of oversupply or undersupply based on company demand. This assessment helped estimate industrial demand for each job type for engineers and technicians by year.

Next, various foreign materials—in particular the classification of vocational training occupations in Japan and related documents published by the International Labour Organization (ILO)—were used to set the scope of vocational training and to organize subjects of training courses. In developing the actual content of vocational training, government invited several experts and professionals to the special "Vocational Training Standards Deliberation Committee." The committee was temporary, without legislative basis, and had about 200 experts and professionals with specializations in certain occupations (with at least three experts assigned to each occupation).

Originally, the experts were supposed to have a certain level of field experience, along with their knowledge of specific occupations. However, most of the commissioned members were university professors and technical high school teachers because Korean industry was only in its initial stage of industrialization at the time, and it was very difficult to find field experts with enough knowledge and experience to develop training standards. Nevertheless, government tried to include at least one expert with industry experience in each working group for one occupation. This allowed industry experts to explain their field experiences to other working group members and reflect them as training standards, even if they lacked the ability to produce official documents.

The vocational training standards commissioned by the labor commissioner on these committee members included those on the content and facility for delivering each type of training course for skilled workers. As a result, the committee was able to develop 139 training standards from the 200 occupations originally targeted. Obviously, the first vocational training standards were not the original ones for the Republic of Korea, as at that time the country did not have not just enough time and budget but also enough expertise to develop its own standards. Therefore, most of the training standards had to be borrowed from Japan, which dominated the country in the past and had a significant impact on the basic framework of the Korean economy. However, they played a

---

[17] The Economic Planning Board played a key role in the early growth of the Republic of Korea by establishing and executing comprehensive plans for economic and social development, coordinating investment plans, and managing budgets. The board was launched on 22 July 1961 but was abolished at the end of 1994 as these government-wide plans were discontinued after the 1990s.

sufficient role in specifying the basic framework and content of the vocational training courses necessary at that juncture of the country's economic development. In other words, the training courses, which were operated irregularly and sporadically without socially unified standards until then, were eventually transformed into a comprehensive system based on consistent standards. In addition, these early efforts contributed to having the supply of skilled workers closer to industrial demand by allowing the training level to be classified according to the level of training content.

Government continued developing content for vocational training, with content standards established and regulated by the Ministry of Health and Social Affairs in 1969 in accordance with the Vocational Training Act, 1967. The act was first revised in 1976, then in 1982, and then in 1991 according to the vocational training development plan. This last revision was noteworthy for allowing increased flexibility in applying training standards; that is, while a long-term training program of more than 1 year still had to comply with training standards, a program lasting less than 1 year need not comply and was even allowed to make modifications as appropriate. In terms of actually developing training standards, government initially relied on individual experts invited to temporary committees. However, by 1984 the government had commissioned the Vocational Training Institute as the agency for training standards development, henceforth institutionalizing the activity.

Prior to the Republic of Korea's enactment of its vocational training law in 1967, the country's vocational training institutes were using translated versions of Japan's vocational training workbooks (Jeong 2008). In the 1970s, the number of vocational training personnel increased rapidly as these institutes were established and the mandatory vocational training system was introduced across enterprises. Demand for vocational training materials thus increased, and in response, the Ministry of Labor (previously referred to as the Labor Office) began to develop training materials using the Vocational Training Promotion Fund. At the request of the Ministry of Labor in 1978, the Labor Welfare Corporation first developed vocational training materials, and since the founding of the vocational training institute in 1980, the institute has been in charge of developing vocational training materials. In 1991, the Vocational Training Institute was incorporated into the Korea Industrial Manpower Management Corporation, which has since assumed the role of developing and publishing vocational training materials.

Discussion

In English-speaking countries, the employer-led sector skills council plays a key role in identifying needed skills in the labor market. However, when industrial development is at its early stage, it is difficult to create such an organization. Since work is simple, companies can manage with low-wage workers and thus do not pay much attention to skills development through long-term, systematic vocational training.

The Republic of Korea was no exception. Instead of asking employers to find out their skills needs in the early stages of economic development, the Korean government tried to understand what these skills needs were by communicating directly with industry experts. Of course, it was not easy to gather experts with knowledge of a particular job and experience in the actual work site. Vocational training standards created were thus imperfect and revisions were inevitable.

The Korean government also became aware of the limitations in developing standards when it initially relied on individual experts but eventually developed vocational training standards through stable research institutes with public sector support later on. This institutionalization and specialization of vocational training ran parallel with the country's economic growth; as the structure of Korean industry advanced, the number of companies deeply interested in skills development also increased. In particular, from the 1990s onward, the need to build a demand-oriented human resource base around these companies has gradually spread. As such, policies for establishing an industry-led skills development system must take into account the degree of industrial maturity and the characteristics of labor demand in a specific country. From this perspective, it would be possible to diagnose the practical needs for the establishment and operation of a sector skills council.

2.2. Economic Development through Aggressive Skills Policies Focusing on Future Skills Demand

Background

To understand the degree of importance of skills development policy in the Republic of Korea's entire economic development process, it is necessary to understand its context at the time. First, 70% of the land on the south of the country's peninsula is mountainous, with insufficient space for farming and scarce mineral resources. Conditions for promoting economic development based on natural resources were almost nonexistent. Second, any limited industrial facilities were destroyed during the Korean War.

To overcome its inherent handicaps, and partly to reflect its Confucian culture, the Korean government was thus forced to focus on nurturing its human resources through education and training.

The shortage of skilled workers was serious, however, as the Republic of Korea began its economic development. The measures that the government used to secure a competent workforce were active investment in the overall education system and a very aggressive vocational training policy.

Progress

The enactment of the Vocational Training Act, 1967 was one such aggressive approach, even when industry was still in its early stages. In 1961, just before the beginning of economic development, the country's industrial structure was premodern, centered on agriculture, and mostly consisted of small businesses. The largest part of industrial production in manufacturing was light industry for daily necessities. In the early stages of economic development, labor-intensive industries were mainly fostered, so many jobs were simple and unskilled. As demand for skilled technical workers was limited, it was possible to meet such demand with unstructured and premodern apprentice training offered by companies or industrial high schools.

However, as the government's economic development projects were promoted and industrial foundations gradually formed, demand for skilled technical workers began to increase. About 3 years after the enactment of the first economic development plan, companies started to complain that factories were not operating properly due to the difficulty of securing enough skilled workers.

Yet, the Korean government was already considering introduction of vocational training from the initial stage of the economic development plan in anticipation of increasing demand for skilled technical workers. According to an EPB report to establish the first 5-year plan for economic development, 299,414 people were in skilled technical occupations as of August 1961, among them, 8,616 experts (2.9%), 11,128 technicians (3.7%), and 279,670 skilled workers (93.4%). This means that the ratio of experts to technicians to skilled workers was 1:1.3:33, significantly differing from the ideal pyramid ratio 1:5:25 expected by the government at the time (Kim and Sung 2005).

Given this situation, the Korean government prepared a projected workforce demand during the First Five-Year Plan. The estimated 280,000 skilled workers in 1961 was increased to 490,000 in 1966, the target year, and an average of 40,000 new workers would be required each year. Forecast demand for technicians increased from 11,000 in 1961 to 97,000 in 1966, or an average of 17,000 would be needed each year.

However, these figures were not a statistical estimate in a strict sense, but rather a reflection of the government's political will and goals. In other words, in the early 1960s, there were very few statistics required to forecast mid- to long-term workforce demand (Yoo et al. 1967), and it was not until the late 1960s that a systematic study of labor supply and demand could be achieved. Therefore, the prospect of workforce demand at that time was not accurate, but rather a targeted number of people that had to be in the workforce to realize planned economic growth.

In early 1962, the EPB established a policy to introduce the vocational training system based on the results of such workforce planning and hastened to introduce a vocational training-related act (Seo 2002). In April 1964, the EPB transferred the responsibility of vocational training from the Ministry of Health and Social Affairs to the Labor Office, which was established in 1963 as an affiliate of the ministry and given the role of managing labor-related policy issues. At the time, the EPB director supervised monthly workforce planning meetings, which allowed for the thorough preparation of the vocational training law. Notably, even at that time when industrialization was very low, the government had already attempted to impose vocational training obligations on companies, foreseeing increasing demand for skilled workers. Specifically, the EPB prepared a draft law allowing the commissioner of the Labor Office to order employers or employers' organizations to establish certified vocational training facilities if deemed necessary for the continued supply of skilled workers. This was intended to strengthen the implementation of vocational training by employers, and in effect served as a mandatory vocational training system. These measures were intended to facilitate the supply of industrial workers at the private level and to increase the efficiency of human resource development.

In fact, the law, which was actually introduced in 1967, did not include clauses about compulsory vocational training, because the Ministry of Commerce and Industry was of the view that government could not impose obligations on companies without sufficient conditions for vocational training. Nevertheless, discussion of compulsory vocational training at the time had significant implications for future development of company-led vocational training. In other words, although the government did not impose vocational training obligations on companies, it decided to provide government loans and tax cuts to companies with training facilities that could conduct vocational training on their own. Through this, a social consensus was formed where companies needed to invest in vocational training themselves. In fact, until then, Korean companies had not felt that they should build their own capacity for training skilled workers, and they had little foresight in this regard. Thus, the 1967 vocational training law—which did not even legislate compulsory vocational training—contributed to the improvement of companies' awareness of vocational training at the time, and laid the foundation for future development of company-led vocational training with the enactment of compulsory training in the 1970s.

The establishment of Kumoh Technical High School is another example of this aggressive skills policy. Kumoh Technical High School was an industrial vocational education institution aimed at fostering technical personnel, and the Ministry of Commerce and Industry played an important role in its establishment. As part of government's electro-industrial promotion plan, the ministry proposed the construction of the Gumi Electronic Industrial Complex in 1969 (Lim 2015). In the process, the lack of skilled workers seriously emerged again as an issue. Accordingly, the ministry asserted that a technical high school had to be built in the Gumi area to train highly skilled workers who could be hired directly into the industrial complex. These assertions reflected complaints that educational content or training conditions in technical high schools did not reach levels required by the industry. In particular, technical education was still neglected relative to the humanities, and excellent students were reluctant to enter technical high schools amid difficulty finding jobs after graduation. The prevailing perception back then was that technical high schools were poorly taken care of by society (Lim 2015).

To help change this perception, the Ministry of Education crafted a policy that designated select technical high schools as specialized industrial schools, which then received generous funding. The Ministry of Education identified and classified industrial high schools into three types: mechanical industry high schools, pilot industry high schools, and general industry high schools, accounting for characteristics of the area where the schools were located. In particular, mechanical industry high schools were established to foster a skilled workforce for the defense industry while raising the skill base of the mechanical industry. Kumoh Technical High School was opened as one of the first four schools designated as mechanical industry high schools for training in precision processing jobs.

At the time, the Ministry of Commerce and Industry noticed that it was not possible to establish the technical high school they wanted with the Korean government's budget alone. Therefore, the ministry came up with an alternative to establish a pilot technical high school under its supervision, using the Japanese government's Korean aid funds. After this alternative was reviewed at the First Korea–Japan Economic Ministers' Meeting in 1967, the governments of the two countries developed detailed plans and began construction in March 1971. The goal was to open the school in 1973 in line with the completion date of the Gumi Electronic Industrial Complex, scheduled for May 1972.

Kumoh Technical High School had two objectives: to cultivate skilled workers desperately needed in industry (especially heavy industry) and to set an example for industrial education, which the government considered ideal. It was founded as Asia's largest industrial high school, and part of the intent was to gradually increase admissions to the school and establish several more schools similar to Kumoh Technical High School.

When it opened in 1973, the Kumoh Technical High School curriculum centered on specialized subjects. The ratio between its academic and professional subjects was 40:60, and the ratio of theory to practice time within the specialized subjects, 30:70. In addition, the school planned to provide a training environment that could realize the highest level of training in the Republic of Korea, which included the use of modern laboratory equipment. The budget for this was overwhelming. For instance, in 1974, about 620 million Korean won (W) was set for the cost of facility expansion for all technical high schools. From 1971 to 1974, however, the Korean government spent W1 billion for Kumoh Technical High School, excluding Japanese government funding support of about W1.2 billion. These figures meant that the budget invested in the construction of Kumoh Technical High School for 1 year approximated the total budget invested in the facility expansion of all vocational high schools. At the same time, the government also made special efforts to develop teachers' competencies. To secure teachers who could provide excellent technical education in an equally excellent practice environment, trainers were dispatched from Japan while Korean teachers were trained in Japan for 6 months.

Special support measures were taken to attract excellent students to Kumoh Technical High School. The government planned to select 300 freshmen from all over the country, with 60 students in each of the 5 departments of mechanical engineering, sheet metal welding, foundry, metalwork, and electronics. These prospective students had to be in the top 10% of their class, and they needed to be endorsed by their respective junior high school principals, who deemed that they could play an active role in the country's high-technology future. Once admitted, these select students had their tuition fees waived, and dorms were provided free of charge. The Ministry of Commerce and Industry also decided to grant qualifications equivalent to a second-class skilled craftsperson upon graduation, and even fully guaranteed employment after graduation.

As the government hoped, admission to Kumoh Technical High School was very competitive, and top-tier students from all over the country applied. In 1973, only 500 out of 1,131 applicants passed the screening, while only 360 were selected as finalists. In 1973, 146 individuals (or 40.5%) graduated with the highest honors among 360 new students, 209 (58.1%) were within the top 5%, and 5 (1.4%) were within the top 5%–10%. The situation was similar in 1974: of the 400 freshmen, 163 had the highest honors, and 221 were within the top 5% of the class. Therefore, according to the President's economic assistant, at that time, Kumoh Technical High School was "a school that only geniuses could enter" (Lim 2015).

Kumoh Technical High School students trained in a high-level, hands-on educational environment, and began to stand out in various contests and qualification exams. The first students in 1973—who completed the basic practice level for only 1 year—won one gold medal, two silver medals, and one bronze medal at the National Skills Olympics in 1974. At the 10th National Skills Olympics in 1975, the school's students made outstanding achievements with 4 gold medals,

6 silver medals, and 1 bronze medal. At the 24th International Skills Olympics held in Busan in 1978, Kumoh Technical High School students earned 4 gold medals and 2 silver medals, greatly contributing to the Republic of Korea's overall victory in the International Skills Olympics for 2 consecutive years (22 gold medals, 6 silver medals, and 3 bronze medals over this period). In addition, most of the school's students passed the level 2 national technical qualification exam prior to graduation, or at the end of the second grade in early cases. For example, 334 out of 346 students (about 97%) in the first term, and 373 out of 380 students (about 98%) in the second term acquired at least one national technical qualification before graduation.

Interestingly, the school's graduates were compared with the graduates of Kyunggi High School—an elite academic high school at the time—in terms of social performance (Ji 2014). Comparing graduates who entered these two schools in 1973 and confirmed through alumni, Kyunggi High School graduates had 153 university professors, accounting for 25.2% of the total, followed by managers in corporations (96, 15.8%), and medical professionals (72, 11.9%). Among Kumoh Technical High School graduates who entered the world of work without going to university, the most common jobs were as managers in corporations (88 alumni, 27.0%), followed by technical experts (67 alumni, 20.6%), and civil servants (35 alumni, 10.7%). The fact that many of Kumoh Technical High School's graduates are not only technical experts but also corporate managers means that they have achieved a significant corporate status after graduation, given that many of the school's students had come from poor socioeconomic backgrounds.

Discussion

What the Korean government has repeatedly emphasized in the country's economic development was that, unlike other countries, it had few natural resources but excellent people. The only way to develop the economy was thus to develop human capabilities. Combining the importance of human resources for economic development with the Confucian emphasis on the value of education, the government made active policies to develop human resources in economic development and in some cases pursued policies that even seemed somewhat reckless.

In hindsight, the main reason those investments were able to achieve significant results was due to the favorable global economic environment; as the Republic of Korea was developing, for example, capitalism around the world was simultaneously growing, which lasted until the early 1970s. However, when this "reckless" policy was decided, no guarantee existed that the economic environment would be favorable, as other developing countries were in a similar situation. Therefore, for the Korean government, aggressive policy decisions on human resource development, such as the establishment of Kumoh Technical High School, were made at great risk.

Clearly, having such aggressive policies actually achieve the desired outcome could have been the result of various and complex factors: the authoritarian government—which at times violently repressed internal dissatisfaction with economic development—the fervent social aspiration to escape poverty, and the existence of technocrats that substantially buoyed the developmental state. In addition, a unique global political and economic situation, which enabled the Republic of Korea to receive active support from foreign countries in the context of regime competition during the Cold War, was an important external factor. There is thus no guarantee that the

Republic of Korea's aggressive skills policy can be replicated in other countries with quite different socioeconomic and political contexts. It is nonetheless still important to consider that the Korean government set up active skills development policies for economic development that closely aligned with the actual demands of economic development.

2.3. Status of Technical and Vocational Education and Training Policy Within the Whole Economic Development Strategy and the Conditions for Successful Integration

Background

In the Republic of Korea's economic development, workforce policy has always been a key area. Each economic development plan included a prediction about how much labor supply would increase during the planning period and how many workers would be needed as the economy developed. And to respond to the increase in skilled workers necessary for economic development, it was always important to properly develop and implement a workforce supply plan. For instance, in establishing the First Five-Year Economic Development Plan in 1962, the Korean government also prepared the First Five-Year Technology Promotion Plan to promote technology development. As part of the plan, there were measures to secure human resources in the technology sector, as cited in the First Workforce Development Plan (1962–1966). The plan classified the technical workforce required for industrialization into three groups, i.e., experts, technicians, and skilled workers, and proposed a road map for supplying them.

Of course, other countries that promoted government-led economic development have often included workforce planning as part of their comprehensive development strategy. Thus, the mere existence of such plans, as part of the development strategy, could not be regarded as definitive evidence proving the high importance the Korean government gave to TVET policies. Rather, given the existence of such plans, a concrete review on active attempts of the government to fill the deficiencies of such plans would be more informative in measuring the importance that the Korean government placed on skills development. In this regard, the following briefly examines how the government approached its workforce development plans in the early stages of economic development, then explains the Technology Manpower Training Plan established in response to an urgent order from then President Park Chung-hee as a concrete example of the high importance of skills development.

Progress

Looking closer at how the government dealt with workforce development in its economic development plan, the Korean government projected in its First Five-Year Economic Development Plan that in 1966 (the target year), labor supply would be at around 11,868,000, or 114.2% of the base year; and that the number of workers would be at around 10,111,000, or 128.3% of the base year. In this regard, the First Five-Year Technology Promotion Plan was established in 1962 to ensure sufficient supply and quality of technical workers for plan implementation. Based on figures from the Korea Technical Human Resource Survey Report prepared in 1961, this plan estimated the number of technical people required and the number of people currently available.

According to the plan, the composition ratio of experts, technicians, and skilled workers in the target year was set at 1:5:25. Training measures and operational plans by skill level were then established for experts and technicians in each field as necessary, to realize these "desirable" composition ratios (Kim and Sung 2005).

Next, the Second Five-Year Plan for Science and Technology Promotion implemented in July 1966 in parallel with the Second Five-Year Economic Development Plan included measures to train skilled workers. The scope of workforce initiatives was expanded because economic development had begun to show a shortage of competently skilled workers in the labor market. As a result, industry–academia cooperation on various vocational training-related measures was emphasized. Specifically, a central vocational training center was established under the Ministry of Labor to carry out various practical support activities related to vocational training. This included training the trainers for vocational education, conducting qualification skills tests, and publishing training materials. Public vocational training centers were also set up in major industrial cities and accredited vocational training centers were established in large companies, to provide training for workers belonging to these enterprises while allowing consignment training for workers of small and medium-sized enterprises.

In line with the establishment of the Third Five-Year Economic Development Plan, workforce development was explicitly separated from science and technology promotion measures, which led to a renaming into the Five-Year Manpower Development Plan. This plan aimed to focus more on improving the quality of vocational training rather than adjusting or increasing its quantity. Public vocational training focused on heavy and chemical industries such as machinery, metals, and shipbuilding, as well as human resources development in the main and export industries. To promote voluntary vocational training in industry, it was also proposed to promote vocational training under the employer's responsibility and deploy a vocational training contribution system for the procurement of investment resources. In addition, measures were taken to ensure a smooth match between workforce supply and demand across the labor market, such as improving the qualifications system, supporting overseas employment of skilled workers, and improving the quality of private employment support services.

Another notable example of the government's holistic approach to workforce development was the Technology Manpower Training Plan established in 1967. According to Seo (2002), the plan was established at around the end of 1966 when the shortage of skilled workers in industrial sites was made manifest. This was the time of the First Five-Year Economic Development Plan and the vocational training law was just decided at the National Assembly. During the First Five-Year Economic Development Plan, industrial foundations such as factory facilities were being prepared as a result of the government's aggressive industrial development policies. During this process, there had been growing complaints that it was becoming increasingly difficult to secure technical personnel, especially skilled craftspeople.

On receiving a report on this situation, then President Park Chung-hee held a meeting of concerned ministries on 5 December 1966 to resolve the issue. At the meeting, he issued an urgent order for government to take measures to cultivate the workforce in technical fields by making full use of all domestic facilities.

The following items were included in the urgent order to the cabinet:

(i) Various military facilities such as maintenance factories, telecommunication schools, logistics schools, vehicle bases, and factories of state-owned and private enterprises should be opened for technical training for nearby non-school youth.
(ii) Given insufficient government finances at the time, without additional government assistance forthcoming, the existing facilities should still be used to the maximum, so that even a small number of people could receive short-term technical training.
(iii) The Ministry of Education was to make basic guidelines for technical training and find ways to officially recognize certificates from the training.
(iv) The minister without a portfolio should gather practitioners of related ministries to prepare detailed methods for technical training.

In response to the president's urgent orders, the minister without a portfolio convened relevant ministries and prepared the Guideline for Fostering Technological Manpower Resources Using Domestic Spare Facilities, with detailed roles and responsibilities for each ministry reflecting the president's prime directives.[18]

At this point, the Guidelines for Nurturing Manpower in the Technology Industry played a significant role.

First, the promotion of this guideline was an opportunity to recognize vocational training courses among related ministries that had not been interested in vocational training. In particular, the president's direct interest in cultivating an industrial workforce made imperative the crafting of vocational training policies. During his tenure, whenever President Park Chung-hee had trips to industrial sites, he rarely missed a chance to visit the local technical school or vocational training facility to encourage activities and listen to their difficulties. After returning from these visits, he instructed public officials to resolve these difficulties and strengthen government support. To lift the pride of skilled workers and emphasize the importance of fostering them, he coined the phrase "Pioneers of Modernization of the Motherland" and resonated it across institutions related to vocational education and training. He also paid special attention to the conditions of vocational training, including the expansion of vocational training facilities and the provision of dormitories for trainees and apartments for teachers.

Second, the guidelines provided an opportunity for the Labor Administration to implement vocational training policies from the beginning of its establishment. Although there were no "training-only" facilities, the guidelines also allowed vocational training programs to expand their operations by utilizing the spare facilities of public and private organizations. In the vocational training plan prepared by the Labor Office in the early days of the establishment, the government

---

[18] In fact, since the training courses covered by this guideline were vocational training courses, it can be said that the head of the Labor Office in charge of vocational training should have taken the lead in preparing this guideline. Nevertheless, the president ordered the minister without a portfolio to develop guidelines because he wanted to show his strong political support to vocational training policies. At the time, the Labor Office was unable to strongly assert the vocational training policy required by the industry because it had a significantly weaker voice in the government than the Ministry of Industry or the Ministry of Defense. Although the minister without a portfolio was responsible for developing guidelines for vocational training, in the actual process, the Labor Office had to have considerable authority and responsibility in practice.

only had licensed trainers' training, supervisory training courses, and support to cultivate craftspeople through training centers for 10 affiliated companies installed in enterprises under the Industrial Education Promotion Act, 1963. Even for training toward a vocational teacher's license, conducted directly by the Labor Administration, there was no self-training facility. As such, before the establishment of the central vocational training center, the meeting room or school classroom of the Korea Chamber of Commerce had to be rented as a training venue. The expansion of vocational training programs by mandating the use of all of the various spare facilities in the Republic of Korea was thus a crucial opportunity for government to emphasize the importance of training skilled workers.

Third, the various participating institutions served as a starting point for vocational training institutes in the Republic of Korea. The Korea Electric Power Company, the Korea National Oil Corporation, *Geumseongsa*, the Korea Shipbuilding Corporation, and Kia Motors were companies that represented the focus on vocational training in business from that point forward.

Discussion

It is clear that TVET policy has always been of high importance in the Korean economic development process. Since almost all developing countries are naturally interested in skills development, it is therefore important to learn from the experience of the Republic of Korea and also have governments scrutinize the conditions under which their respective TVET plans can actually be formulated and implemented.

Vocational training policies in the Republic of Korea in its economic development were successfully designed and promoted by combining a number of conditions: (i) strong leadership in economic development within an authoritarian government, (ii) dedication of young civil servants in economic development and vocational training, and (iii) effective advice and support from foreign countries.

(i) *Strong leadership within an authoritarian government.* The Republic of Korea had an authoritarian government that centered on President Park Chung-hee, who made economic development a top priority and suppressed political opposition. Though in this context human and labor rights could be neglected, it did allow for very high efficiency in government policy development and promotion.

(ii) *Dedication of young civil servants in economic development and vocational training.* Competent officials were also highly important. The government's policy-making process has traditionally been led by professional bureaucracies, going as far back as the *Gwageo* system introduced in the early 10th century. Over time, it has become a tradition to select officials through an extremely strict government-managed examination system and ensure their stable work. This tradition continued despite institutional changes during Japanese imperialism, and even after liberation, young and smart youth were selected through rigorous testing and assigned to government affairs. At that time, the only good jobs were in government, and those who were selected competitively felt a strong sense of responsibility and pride that they were leading the country toward progress and embodying government's catchphrase of "modernization of homeland." They maintained an open attitude when consulting domestic and foreign experts in related fields, studied other country cases, and actively recommended innovations that seemed relevant to TVET in the Republic of Korea. These recommendations went through a hierarchical process within the bureaucracy that involved consultation with related ministries and the president himself, who would translate these to policy once convinced of their relevance to the goal of economic growth. Such smooth policy making was only possible through authoritarianism, which did not leave any space for political opposition parties and trade unions.

(iii) *Advice and support from foreign experts.* TVET specialists from other countries also helped in forming the country's vocational training policies. Among these specialists was Albert Schnitter, a German expert dispatched to the Labor Office in the late 1960s who advised the Republic of Korea for more than 10 years.[19] He graduated from the Institute of Technology after completing apprenticeship training in Germany and had extensive theoretical and practical expertise. At that time, young civil servants interacted almost daily with German advisors, seeking specific alternatives to develop vocational training in the country. Also, in the late 1970s and early 1980s, the ILO sent representatives from a number of developing countries to the Republic of Korea to share state-of-the-art in vocational training delivery.

It can thus be said that the success of the Republic of Korea's vocational training policy resulted from a combination of internal and external conditions: the Korean government's competent and enthusiastic civil servants, and expert policy advice from specialists abroad.

2.4. Pursuit of Public–Private Partnerships and the Role of Government in Driving Such Cooperation

Background

In general, it is known that education and training have high economic externality; that is, the return on investment for the whole society is greater than the return on investment for the individual. Thus, it is highly unlikely that a socially necessary level of investment will be made if the investment decision is made purely by individuals. Accordingly, the government invests in education and training to ensure socially desirable education and training is achieved.

The problem is that while developing countries need active education and training for their development, the governments of these countries do not have enough resources to do so. As such, developing country governments try to obtain foreign aid and private resources. In the early stages of economic development, the Republic of Korea actively sourced foreign aid to compensate for the lack of education and training finances and tried to entice the private sector to invest in education and training. Skills development through vocational education and training was no exception. The Korean government has implemented various measures to promote private sector participation in vocational education and training, which are reviewed in the following section.

---

19 Interview with experts involved in vocational training at the time.

Progress

After the Republic of Korea was liberated from Japanese imperialism in 1945 and until the late 1950s, the Korean government invested limited government resources in realizing compulsory primary education, while private schools were relied on to promote secondary and higher education. The proportion of private middle schools increased from 44.4% in 1965 to 48.6% in 1970, and the proportion of private high schools accounted for 50.7% in 1965 (Paik 2013).

Overall, from the mid-1950s, policies that had actively utilized private schools as a major pillar of the Korean education system contributed significantly to providing opportunities in response to increasing demand for education. It also allowed for equal access to education, regardless of gender, residential area, and socioeconomic background.

Private schools also expanded into vocational education, albeit slowly because vocational education entailed greater costs than general education. But it was clear that this expansion greatly contributed to the sector's growth. As of 1965, private schools accounted for 54% of all high schools and 35% of vocational education schools. In addition, the proportion of private high school students among all high school students reached 59% in general schools and 39% in vocational education schools.

Government provided incentives such as tax cuts or financial subsidies to encourage and promote the establishment of private schools. First, after liberation, the land reform policy in 1949 had a significant impact on the large increase in private schools. At that time, land reform was to redistribute the farmlands of large landowners—who had accumulated large tracts during the Japanese colonial period—to small farmers. Aiming to increase educational infrastructure through the voluntary investment of large landowners, government decided to exclude farmland for use as school sites from redistribution. The land reform policy thus encouraged large landowners to voluntarily establish and operate private schools by acknowledging them as contributors to social development.

Second, once private schools were established and started to operate, the government provided them the following types of financial aid:

(i) *Support for teacher salaries and operating costs in private secondary schools.* The size of the subsidy was determined based on the difference between the "standard financial demand" by the school level and size set by the government, and the "standard financial income" calculated from the student tuition and school foundation transfer. Since the financial capacity of school foundations had been generally very weak, almost all teacher salaries in private secondary schools were financed by the government.

(ii) *Subsidies to private universities.* The government provided subsidies to strengthen the research activities and educational capabilities of private universities. The amount of government-funded grants was determined based on criteria set by various government projects.

(iii) *Tax benefits for private schools.* In principle, all taxes directly related to educational activities of private schools, such as corporate tax, VAT, property tax, and domestic tax or local tax, were exempted. For-profit businesses owned by the school foundation had to pay the applicable taxes, but at lower rates than usual. In addition, when individuals or organizations donated to private schools, they could receive tax credits as donors to educational facilities.

On the other hand, the government was applying regulations along with the expansion of private schools. In particular, through the enactment of the Private School Act, 1963, the supervisory authority over private schools was strengthened. This law promoted the healthy development of private schools by ensuring their administrative autonomy and securing the publicity of private schools as educational institutions. Specifically, by citing conditions and regulations for establishment and operation of school foundations and private schools, they had a pivotal role in the national education system, but their accountability and operational responsibility were also made clear.

According to this law, the school foundation had to report matters related to financial or personnel management to the supervisory authority for its approval. Through this law, the government tried to ensure that only competent and sound school foundations could establish and operate private schools. An independent school account was created separately from the foundation account so that student tuition fees could only be used for educational purposes, thus helping prevent accounting fraud or embezzlement.

In private vocational training—which began in the mid-1970s—private resources were widely mobilized for vocational training (Jeong 2008; Choi, Oh, and Choi 2009). Previously, the main channels of vocational training were public institutions managed by the government and in-house training facilities owned by enterprises. As demand for vocational training increased, however, a shortage of training supply capacity in government training institutions and corporate training facilities became more evident. In response, the government enacted the Basic Act on Vocational Training, 1976 to approve select training provision by institutes established and operated by private foundations as "accredited vocational training," and regulated by the Labor Administration. Accordingly, vocational training was classified into three categories: public vocational training, vocational training in business, and accredited vocational training.

In 1977, the country had 12 training institutions belonging to social welfare foundations and 21 training institutions belonging to nonprofit foundations. However, the number increased rapidly through a series of deregulation measures for private providers; for example, in 1981 individuals were allowed to establish and operate accredited institutions to encourage participation of private actors in the provision of vocational training programs. After this concession, the number of recognized vocational training institutions increased, from 33 in 1977 to 106 in 1991; this also improved the number of trainees, from 9,817 to 25,190 for the same period. In particular, government support for accredited vocational training has expanded since 1993, and the number of accredited vocational training institutions exceeded 130 by the mid-1990s.

The government continued to push for deregulation to develop private vocational training institutions. For example, when the Basic Act on Vocational Training, 1976 was amended in 1993, provisions were added to protect and foster private vocational training corporations. As a result, restrictions on the entrustment of employer training to private training institutions were eliminated. Originally, only private vocational training institutions with more than 1 year of training performance were allowed to receive entrusted training from companies. After deregulation, private training institutions were allowed to receive corporate training programs right after establishment. In addition, the scope of training that these private vocational training corporations could provide was greatly expanded, allowing training for all except 16 jobs that the government deemed necessary to provide workers for through public training institutions.

However, some regulations remained in place. For a private foundation to be recognized as a training institution, more than W300 million had to be declared as its basic assets, and at least one of its corporate directors had to have at least 10 years of education and training experience. Under these conditions, after obtaining permission to establish a vocational training corporation, it also needed to have facilities, equipment, teachers, and teaching materials to the vocational training standards prescribed in the Basic Act on Vocational Training, 1976.

Individuals who want to establish and operate an accredited training institution had to have at least 5 years of experience in training after receiving a vocational training teacher's license from the minister of labor. They are also required to have facilities, equipment, teachers, and teaching materials that meet government standards prior to recognition as a vocational training institution.

Private training institutes were responsible for training occupations that were not in the charge of public vocational training institutes and in-business training institutes, as well as occupations that were particularly needed in the region or industry. At that time, public training institutes were focused on training for mechanical and special jobs required for the heavy and chemical industries, while in-business training was for building the skills of workers required by the companies themselves. Therefore, private training institutes played a significant role in responding to various civilian training demands, which could not be handled by public training or in-business training.

Discussion

The Korean government has very aggressively attracted private investment to address the lack of financial resources for education and training. In particular, the exclusion of potential school sites from land reform and having the founder of a private school as the chairperson of a private school foundation acted as significant incentives for those who had more than enough land to establish private schools. Considering that it was difficult for private investors to continue paying the operating costs of education and training institutions, government committed to shoulder these expenses as an added form of support.

This encouragement of private participation by government helped expand access for vocational education and training providers and increased the diversity of education and training programs. As the number of private vocational education institutions increased in response, vocational education and training opportunities were expanded for both youth and adults. Although government required these institutions to comply with education and training standards set by the state as a condition for continued public subsidies, government also allowed these institutions to have considerable operational autonomy. Accordingly, these institutions were able to conduct differentiated education and training tailored to their purpose.

However, there is a lot of controversy over whether the quality of education and training of private institutions was adequate. Although government subsidized the operating costs of these institutions, the amounts were not sufficient. In addition, there had been cases where the founders of some private training institutions misappropriated their subsidies, either using the funds to set up other education and training institutions or diverting these into personal accounts.

Though government deployed measures to strengthen the supervision of private education and training institutions to ensure that public funds were used for their original purpose, some erring founders were condoned for their misconduct. This led other private training providers to complain that policies were being selectively enforced, rendering regulation ineffective.

Therefore, the experience of expansion and operation of private education and training institutions in the Republic of Korea shows a fundamental problem of TVET in developing countries. When private resources are mobilized to expand opportunities for vocational education and training, there is a dilemma that, while appealing to the goodwill of private investors, it is impossible to rely on such goodwill unconditionally. On one hand, it is necessary to offer generous incentives to make vocational education and training provision attractive to the private sector. On the other hand, an appropriate regulatory mix is imperative to prevent private school owners from pursuing their own interests at the expense of quality education and training delivery.

2.5. Policies and Measures for Promoting Employer-Led Training

Background

In the early 1970s, the Korean government started to enforce needed measures to prepare for a shift in the drivers of economic growth; that is, from light industry to heavy industry. In fact, industrial policy toward the heavy and chemical industries was not an entirely a new approach, as the establishment of steel and petrochemical plants was included in the Second Five-Year Economic Development Plan, 1967–1971. However, the heavy and chemical industrialization policy announced in 1973 was quite ambitious, considering the circumstances the Korean economy was facing at the time. The government aimed to increase the proportion of the heavy and chemical industry in manufacturing value added, from 35% in 1972 to 51% by 1981, and the share of heavy and chemical industry exports in total exports from 27% in 1972 to 65% by 1981.

The logic behind the government's initiative in doubling the share of the heavy and chemical industries in less than a decade was not purely an economic calculation. On top of severe competition expected for light industries with other Asian countries, nurturing the defense industry to prepare for the possible withdrawal of the United States Forces from the Republic of Korea after the end of the Viet Nam War was the key political factor in promoting heavy and chemical industries during the period.

The government's active leadership was thus crucial in achieving the targeted change. To overcome limited domestic demand for heavy industry and chemical products and achieve economic plan targets, the government focused on export promotion to foster the heavy and chemical industries from an early stage. It then set up a detailed development plan for six industries: steel, nonferrous metals, machinery, shipbuilding, electronics, and chemicals, and intervened in selecting private companies to conduct business by providing incentives such as tax exemptions and financing modalities. Accordingly, heavy and chemical industrialization became the core factor of the economic development plan from 1973.

Progress

With regard to vocational training policy, noting that existing self-regulatory vocational training policy as embodied in the Vocational Training Act, 1967 would not successfully supply a skilled workforce for heavy chemical industries, the Korean government decided to divert to a government-led and regulation-oriented vocational training policy. Specifically, the limits of Vocational Training Act, 1967 became evident in the early 1970s when public subsidies for vocational training were reduced, as government was tightening its budget due to increased expenditures for economic development projects. Such a reduction relied on the assumption that companies would be able to train skilled workers on their own. However, after the subsidy reduction, the number of in-plant trainees decreased to 7,999 in 1973, which caused a serious mismatch in the supply and demand of skilled workers and resulted in an intense scouting war for skilled workers among companies. For example, in some companies, 30% of all workers transferred to a competitor company each month, enticed by a significant wage increase.

The Korean government realized the urgency to address such a skills mismatch by deciding to impose a training obligation on companies. The Vocational Training Special Measures Act, 1974 was first prepared in August 1974, passed without any modification or objection at the regular session of the National Assembly in December 1974, and forced on companies 4 months after ratification. Such a hurried implementation of vocational training obligations, without sufficient social reviews, put a considerable burden on companies (Seo 2002). Nonetheless, its basic structural features persisted for more than 2 decades.

The act obliged companies with 500 or more workers in six industries—mining; manufacturing; electricity, gas, and water supply; construction; logistics and telecommunications; and services[20]— to provide in-plant vocational training for new recruits during the Third Five-Year Economic Development Plan, 1972–1976. In 1976, the government strengthened the compulsory in-plant training system even further by enacting the Basic Vocational Training Act, 1976. The size of companies obliged to provide in-plant training under this new act used to be large companies having more than 500 employees only. However, it was further expanded to include medium-sized companies with more than 300 employees. The number of trainees was set by the government

---

20 Although the act stipulated that all service industries were subject to the vocational training obligation, it exempted the service industry from training obligations except for "repair work." The laundry industry was added in the list in 1987 to provide training but only for companies with 200+ workers. Therefore, in contrast to the construction and manufacturing industry, most service companies were exempted from training obligations.

annually, based on demand for skilled workers in each industry, but not exceeding 10% of the total number of workers in companies. In the same year, the Vocational Training Promotion Fund Act, 1976 was promulgated to effectively utilize funds from vocational training levies. To promote vocational training, the funds could be used to pay for entrusted training expenses, training allowances, accident compensation, subsidies for public training facilities and in-plant vocational training, survey and research projects, awareness and publicity activities, and so on.

Such a mandatory in-plant vocational training system contributed significantly in supplying adequate labor for heavy and chemical industries and achieving economic growth. Although private companies strongly resisted the system because of its inflexible, uniform, and unrealistic regulations, most companies abided by the vocational training obligations. For example, in 1978, about 70% of firms with training obligations introduced in-plant vocational training programs while the rest chose to pay vocational training levies. As a result, the number of skilled workers trained through in-plant vocational training steadily increased, from 10,799 in 1972 to 42,667 in 1975 and 96,820 in 1976.

Discussion

The implementation of the vocational training system during this period substantially contributed to providing skilled workers for industrialization and raising people's awareness of the need for and importance of practical skills, especially among those who favored general education over vocational education and training. With active government intervention, the foundation for the vocational training market was beginning to form: the number of qualified skilled workers as well as vocational training centers and authorized vocational training institutes increased noticeably. Considering that the Republic of Korea's per capita income was less than $1,000 until 1977, its success in implementing the vocational training system was quite impressive. The introduction and settlement of compulsory vocational training in the 1970s clearly reflects the unique political and economic environments of the Republic of Korea.

As a developmental state (Ashton et al. 1999), the Korean government had a strong influence on almost every aspect of economic development. The Blue House under the leadership of President Park Chung-hee and the EPB were deeply engaged in the economic decisions of the private sector, from resource allocation to even labor disputes among individual companies. With the core objective of promoting domestic companies, government provided favorable economic and industrial conditions to large companies in return for nurturing industrial workers. For example, in August 1972, just 2 years before the introduction of the compulsory vocational training system, a special measure called the "3rd August Measure" was taken to ease the burden of private debt repayment on companies. According to the measure, all companies' private debt repayments were postponed for 3 years, with the interest rate lowered by two-thirds, relieving the burden of debt immediately. Having such special relationship with companies, government was able to legislate the compulsory vocational training system in a short time with little consultation with companies.

Moreover, domestic politics had a significant impact on introducing the new vocational system without much resistance. President Park Chung-hee, deeply concerned about the possibility of US troops withdrawing from the Republic of Korea after the end of the Viet Nam War, strongly pursued political stability to gain advantage over the Democratic People's Republic of Korea in the event of a military confrontation. Accordingly, in October 1972, he declared emergency measures that included the dissolution of the National Assembly and established the Yushin Constitution, which strengthened the presidency by filling one-third of the members of the National Assembly with the president's recommendation and eliminating restrictions on the reappointment of the president. Until President Park Chung-hee's assassination in 1979, the National Assembly and civil society were banned from freely expressing political opinions or criticizing the government's policies in public. All government policies were implemented without much resistance, which was favorable for the early settlement of the compulsory vocational training system that could be regarded as a serious burden to companies.

2.6. Contributions of the Qualifications System Despite the Absence of a Comprehensive Qualifications Framework for All Education and Training Programs

Background

For the supply and demand of labor in the labor market to match smoothly, accurate information about competencies is needed for jobs and for those seeking work. Qualifications, on one hand, indicate the competencies each job requires, and on the other, prove that those with qualifications have those needed skills. Therefore, many countries have a high interest in systematic management of qualifications by profession. However, as the number of jobs requiring professional qualifications increases in the labor market, the relationship between qualifications becomes complex, and in some cases, confusion may arise. To solve this situation, a qualifications framework that comprehensively displays the interrelationships of various vocational qualifications, including those with academic degrees, is needed. In this respect, the necessity of establishing a national qualifications framework has been strongly pointed out in the recent English-speaking advanced countries. The Korean government is also considering establishing a national qualifications framework that includes academic degrees and vocational qualifications.

However, in the Republic of Korea, various vocational qualifications were created and in place in the process of economic development, even though the national qualifications framework, as discussed, until recently did not exist. Efforts to reshape the interrelationships among frameworks continued as various qualifications emerged, but no attempts were made to reshape the relationship between all kinds of degrees and qualifications across the entire labor market. How could the qualifications system develop in line with industrial demand in such a situation? The following looks at how the qualifications system in the Republic of Korea was formed and developed in the absence of a national qualifications framework. Through this, it will be possible to infer the practical policy implications of the national qualifications framework in the process of economic development in developing countries.

Progress

In the Republic of Korea, as economic development began in earnest in the early 1960s, the number of professional qualifications began to increase. Until then, while agriculture was still an important source of jobs, the labor market was not specialized by occupation, and few jobs required professional qualifications. However, as the economy gradually developed and new types of occupations (mainly in manufacturing) appeared, the demand for professional occupational qualifications increased.

Accordingly, in the 1960s, new professional qualifications began to appear in various occupations by imitating foreign cases (Human Resources Development Service of Korea 2002, Seo 2002, Jeong 2008). Yet the government remained in charge of setting standards for professional qualifications and issuing actual certificates, amid concerns that the creation and issuance of professional qualifications by the private sector could have several consequences. In those years, about 10 ministries operated 26 professional qualifications systems individually for each unique purpose according to 19 laws and regulations, such as the Professional Engineers Act, 1963 and the Construction Industry Act, 1958.

However, several problems appeared in the process of each ministry running professional qualifications for each purpose. There were cases where the content was similar between different qualifications, or the criteria for classifying the qualification levels were inconsistent. The overlap in qualification categories and the inconsistency in qualification levels confused those who wanted qualifications and had a negative impact on the actual labor market utilization of qualifications.

The Korean government recognized the necessity for consistent management of qualifications and proposed several institutional changes. First, the Vocational Training Act, 1967 stipulated that the "training of workers is required for manufacturing and other industries by conducting vocational training and skill tests for workers." Through these regulations, the Labor Office took responsibility for managing the qualification test. At the time, government decided to grant qualifications through skills tests despite its limited oversight capabilities; that is, skills tests for applicants were the most cost-effective and feasible option for the Korean government, which did not have sufficient capacity for monitoring the actual training courses of individual training institutions. The actual skills test for professional qualifications could be conducted by the Labor Office itself or entrusted to local governments or vocational training institutions.

Operationally, for the skills test under the Vocational Training Act, the first functional test based on the integrated management system for 15 qualifications was conducted in 1968 and divided into written and practical portions. Part of the practical test for some qualifications was entrusted to the Korea Shipbuilding Corporation and the Hangul Mechanization Research Institute. There have been various developments since then, such as the preparation of test standards by qualification level and improvements in test methods. A total of 122,000 successful applicants from 1,999 occupations were produced through the integrated functional skills test under the Vocational Training Act, which continued to operate until the enactment of the National Technical Qualification Act, 1973. At the time, the integrated functional skills test under the Vocational Training Act greatly contributed to unifying the qualifications system for professional qualifications that were distributed and operated by government departments.

However, until the early 1970s, government ministries were still in charge of managing various professional qualifications, and the problems noted earlier continued. First, as the details of each qualification were confusing and even inconsistent with each other, it was difficult to apply the principle that people with the same skill level and skill content should have the same qualifications. Next, as similar qualifications continued to operate separately, those who wished to work in fields to which these qualifications were relevant had to apply and acquire similar qualifications several times. In addition, due to the internal confusion in the qualifications system, the relationship between the qualifications system and the vocational education and training system was not appropriately established.

To address these problems, the introduction of an integrated qualifications system in which the state granted professional qualifications according to a uniform standard for technicians and skilled workers was actively considered. The Korean government thus enacted the National Technical Qualifications Act in 1973, consolidating various professional qualifications regulated under various laws into one statute, and abolishing some qualifications if found redundant or overlapping with others. To maximize the contribution to economic development, qualifications related to heavy and chemical industrial technologies such as machinery, metals, and chemicals, were first reorganized in the process of enacting the National Technical Qualification Act. In addition, the degree of relationship between vocational education and training and professional qualifications, and the possibility of implementing a reliable skills test, were issues considered important in the improvement of the professional qualifications system.

As a result, 128 qualifications corresponding to the 19 laws and regulations of each ministry were absorbed into the national technical qualifications system. Such a system also stipulated a grading system by qualification level for each technical field and skilled worker field: that is, from high to low, for technical field; professional engineer—engineer 1st grade—engineer 2nd grade; and for skilled worker field, master craftsperson—craftsperson 1st grade—craftsperson 2nd grade—and assistant craftsperson. For the qualifications for technical fields, a specific division of qualifications for engineers also reflected the divisions of departments in universities, especially engineering departments. Therefore, students enrolled in engineering departments were able to obtain not only a bachelor's degree but also an engineer 1st grade qualification by passing a certification exam when graduating from university. The detailed qualifications of the skilled worker field were set by reflecting the demand for skilled labor that emerged during the period of economic growth. For each grade of the national technical qualification, test-eligible requirements were set and the applicants had to pass a three-stage skills test—including a written test, a practical test, and an order to obtain qualification. In specifying the test-eligible requirements for each grade, the academic background and actual work experiences were comprehensively considered.

Those who had obtained a professional qualification before the enactment of the National Technical Qualification Act, 1973 were recognized as those who obtained a technical qualification under that act. To increase social acceptance of national technical qualifications, preferential measures for those with national technical qualifications were adopted and strengthened. Subsequently, it was stipulated that national technical qualification holders could receive privileged treatment in employment in the company, such as higher salary and better chance of promotion.

In 1976, the Korea Technical Validation Authority was established as an agency dedicated to the technical qualification examination in accordance with the National Technical Qualification Act. It was responsible for almost every task necessary for the implementation of a national technical qualifications system: development of skills test items, actual management of skills test implementation, issuance of certificates and their registration, as well as research and development. In 1977, the agency conducted the national qualification test for more than 100,000 applicants on 199 qualifications. Through these activities, the agency made a significant contribution to the development of the professional qualifications system in the country.

In this process, the Korean government created, modified, and abolished qualifications from time to time according to the specific needs of individual occupations or industries without a comprehensive qualifications framework, widely known as the National Qualifications Framework, which indicates the compatibility between various academic degrees and qualifications across the entire labor market. As it was promoting economic development, the Korean government did not know of such a framework itself, and even if it did, it was unlikely that it would actually craft it, given its limited policy capacity then. As a result, the Ministry of Labor continued to focus on the vocational qualifications necessary for industrialization and tried to manage them within the framework of the national technical qualifications system, paying limited attention to academic compatibility. On the other hand, some qualifications were still managed by individual central ministries. Therefore, even though the Republic of Korea's vocational and technical qualifications system contributed greatly to fostering the necessary workforce for economic growth, it cannot be denied that the qualifications system had shortcomings, in particular, nonconformity or disconnection from the academic sector, and fragmentation within the technical qualifications. In this regard, the government came to consider the establishment of a National Qualifications Framework following the related experiences of advanced English-speaking countries.

Discussion

Qualifications summarize the competencies required by a job on one hand, and represent the competencies a person has on the other. The qualifications system is therefore an important institutional instrument for the development of competency through education and training, and for matching supply and demand in the labor market. However, in the early days of the Republic of Korea's economic development, even qualifications were not a decisive factor for entry into the labor market, except for qualifications with the nature of a license, such as for teaching or medical practice. This was due to two reasons. First, in the early stages of economic development, most labor demand was for unskilled labor. Since they did not require special qualifications, many unskilled workers were able to find work without qualifications. Second, in the Republic of Korea, academic education played a decisive role in determining the position of a person in the labor market. Traditionally, jobs were ranked in the order of white collar–agriculture–industrial–commercial, and in particular, for jobs with high social status, their level, and the school they graduated from were considered key employment factors. Further, in the process of industrialization, companies used to recruit inexperienced people through a large-scale public recruitment process and train them in-house to cultivate needed competencies.

The social necessity of qualifications was therefore not very high, at least in the early stages of economic development, except for occupations that the government needed to regulate for hygiene or health issues. As industrialization progressed, the need for specialized technical qualifications increased and, accordingly, the Korean government created and managed various qualifications. Even so, the government did not establish and operate a comprehensive qualifications framework that encompassed both vocational qualifications and degrees. Nevertheless, government did not have much difficulty supplying the labor market with the staff required for industrialization through the rather fragmented vocational qualifications system. This experience indicates that a very comprehensive qualifications framework may not be indispensable in operating a qualifications system according to the needs of the labor market.

Actually, the lack of a comprehensive framework to anchor the country's vocational qualifications system had both advantages and disadvantages. It was possible to respond very flexibly to changing labor market demands through changes in individual qualifications. That is, when the need for a new qualification was raised in a new growth industry, the government was able to create and apply a new qualification standard almost immediately. On the other hand, confusion and even conflicts emerged among various qualifications. In particular, there were instances in which companies and individuals were confused about qualifications with similar names but different actual content, or vice versa. In this situation, a comprehensive qualifications framework would have alleviated such confusion.

However, given that the Korean government at that time did not have enough human and material resources to operate a comprehensive qualifications framework, such confusion could be regarded as inevitable to some extent. In any case, the advantages of a comprehensive qualifications framework were obvious. Considering the substantial resources to create and manage such a qualifications framework, however, it would be more desirable to judge the necessity of a qualifications framework by considering the features of the labor market, such as its industrial maturity or the dominant allocation mechanism. More specifically, in the labor market of many developing countries, occupations are not very diverse, and each occupation does not require a considerable amount of education and training. Therefore, while the significance of a national qualifications framework itself cannot be denied, careful examination should be conducted to assess the extent such a framework is really needed. In the end, the judgment on the necessity of the national qualifications framework should be based on a detailed evaluation of the economic and labor market conditions of each country.

2.7. Relationship Between the Development of General Education and Technical and Vocational Education and Training

Background

Vocational education and training is greatly influenced by the overall development of the education and training system of which it is part, and is achieved through multiple channels. First, vocational education and training is conducted after primary education, and is affected by how much basic competencies are acquired through primary or lower secondary education. If many students do not have basic literacy or numeracy through general education, vocational education and training that builds professional skills for each job will be difficult. In this regard, the proper development of general education at the elementary or lower secondary level is a prerequisite for the development of vocational education and training.

Next, TVET may form a competitive relationship with other education and training programs, such as upper secondary general education or higher education. That is, vocational education and training may or may not be attractive, depending on the types of job opportunities available to TVET-trained youth compared to youth who have undergone other education and training programs. It is clear that job opportunities for TVET-trained youth are fundamentally affected by the degree of development of industry. But employers also adjust the direction and targets of hiring personnel, even with the similar level of labor demand, through consideration of the relative supply of those who have undergone similar education and training.

The following examines how TVET in the Republic of Korea was affected by the overall education and training system, especially the development of primary and secondary education, followed by the development of higher education. Through an understanding of this process, it would be possible to appreciate the complex relationship between vocational education and training and other education and training types—complementary on one hand, and competitive on the other—in the context of changes in industrial demand.

Progress

Aspiration for education began to surge as soon as the Republic of Korea was liberated from Japanese imperialism. During 1945–1948, while under US military rule, the number of elementary school children increased from 1.36 million to 2.66 million (Yoon et al. 2012). In 1948, government established compulsory elementary education as one of its urgent tasks, but the outbreak of the Korean War in 1950 saw school enrollment drop from 74.8% in 1948 to 69.8% in 1951. When the war ended in 1954, the government crafted its Five-Year Compulsory Education Plan, which aimed to raise the enrollment rate of elementary school-aged children to 96% by 1959. However, the actual number of students increased faster than expected, from 2.68 million in 1954 to 3.56 million in 1959, exceeding the original target enrollment rate by 0.4%. This increase was remarkable considering the massive destruction of facilities and subsequent economic difficulties as a result of the Korean War. Government was equally invested in this education mission, and by 1959, compulsory education expenses accounted for nearly 81% of the total budget of the Ministry of Education. The funding share of education in the total government budget also increased from 4.0% in 1954 to 9.2% in 1957 and 14.9% in 1960, and the share of the education budget to gross national product increased to 2.5% in 1960.

Yoon et al. (2012) assessed that the rapid universalization of elementary education played a crucial role in the postwar economic development of the Republic of Korea. Along with other factors such as the country's homogenous population, monolingualism, and aid from the international community, universal elementary education helped equip the youth with basic competencies to enter the labor market.

However, the gains from prioritizing education were not immediate. The Korean economy in the 1950s struggled with per capita incomes at below $80, making it one of the world's poorest countries. Under these circumstances, ensuring that school facilities cater to every student was not a possibility. The double-shift school system combined with overcrowded classrooms were the official norm: statistics showed that more than 80% of all elementary school classes were operating on at least a double shift with some schools even running triple shifts, with students per class at about 70 to 80—and as high as 90 students in major cities. An elementary school in Seoul, the capital, was reported to have a total of 120 classes serving 8,000 students (or close to 67 per class). Clearly, the government's approach at the time was to educate as many children as it can while setting aside—at least temporarily—concerns about education quality. There were attempts to address quality issues by increasing the number of schools and classes and hiring more teachers, but in a situation where social aspiration for more educational opportunities was overwhelming, priority had to be placed on expanding those opportunities.

After primary education was made universal in the 1950s, the rapid expansion of secondary education followed a similar pattern. The number of middle school students—around 400,000 in the mid-1950s—rose sharply in the 1960s: it surpassed 600,000 in 1961, was at 900,000 in 1970, and at 1.5 million in 1975.

Figure 3 shows this advance of student enrollment by education level since the mid-1960s. Along with the expansion of education, vocational training for youth who had graduated from junior high school but did not enter senior high school also started in earnest from the mid-1960s. By going through all these levels of education and training, competent young workers became prepared to enter the labor market, a key success factor in the country's economic growth.

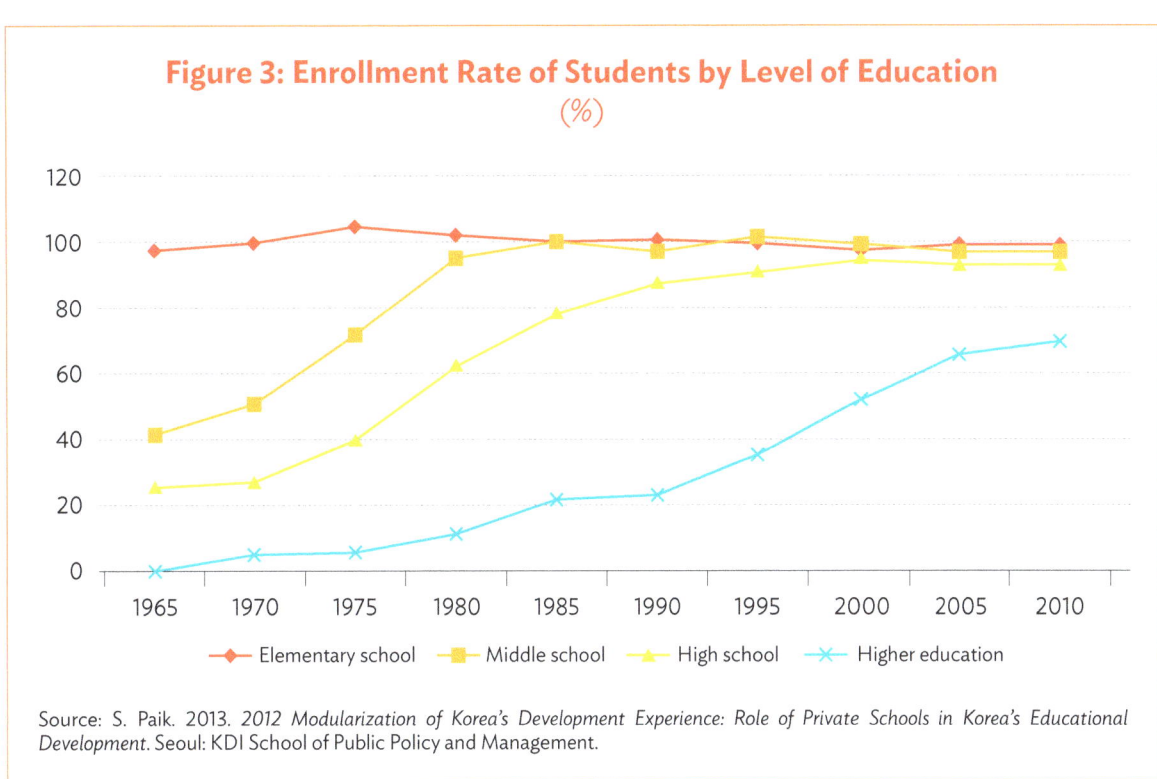

Source: S. Paik. 2013. *2012 Modularization of Korea's Development Experience: Role of Private Schools in Korea's Educational Development*. Seoul: KDI School of Public Policy and Management.

This fast-paced expansion of elementary and lower secondary education has reduced illiteracy among adolescents to almost zero. The Korea National Statistical Office even suspended statistical surveys on the illiterate population from 1966. Worth noting, however, is that by itself, knowing *Hangul* (the Korean alphabet)—which is very easy to learn—does not lower illiteracy. That is, according to Kim (1999), about 78% of Koreans were illiterate based on the 1930 census report conducted by the Japanese governor-general, when the language of education was in Japanese. Although knowing *Hangul* is clearly effective in reducing illiteracy, education is essential to eliminating illiteracy. As such, compulsory elementary education in the 1950s followed by the expansion of lower secondary education significantly improved literacy among youth as they entered the labor market.

Meanwhile, enrollment in higher education was slightly above 20% in the early 1990s but was at around 70% in the latter half of the 2000s. While development in education had undeniably been induced by changes in labor demand due to structural changes in Korean industry, these also reflected the increased social aspirations for higher education and the demand for deregulation of higher education institutions. This growth of higher education had a significant impact on the relative attractiveness of vocational education and training: when opportunities for higher education were limited, youth in vocational education and training had the advantage in obtaining low-ranking white-collar jobs or higher-level blue-collar jobs. Therefore, among students who graduated from middle school at the time, those who could afford it chose to go to college through general high school, while those who were smart but had financing issues chose to go to vocational high school.

In fact, three of the country's past presidents graduated from vocational high schools: President Kim Dae-jung (1998–2003) studied at Mokpo Commercial High School when it was difficult for Koreans to attend general high school during the Japanese colonial period. President Roh Moo-hyun (2003–2008) attended Busan Commercial High School in 1966, and President Lee Myung-bak (2008–2013) went to Dongji Commercial High School in 1960. The fact that three presidents graduated from vocational high schools shows that a number of poor but smart students went on to these schools when opportunities for higher education were limited. However, the rapid expansion of higher education made vocational education and training relatively less attractive.

This effect can be seen in the decreasing wage premium of vocational school graduates. According to Choi and Choi (2004), until the early 1990s, vocational education graduates received 10% to 20% higher wages than general high school graduates (provided that all individual circumstances such as gender, years of education, and experience remained the same). However, starting in 1995, this began to change as vocational education graduates started to earn less than their high school counterparts. Along with the decreasing attractiveness of vocational education, the share of vocational education students also started to decrease. That is, in the 1960s and 1970s, when opportunities for higher education were limited, between 40% and 50% of students were in vocational education. Even in the 1980s and 1990s, when the economy was industrializing and upper secondary education was expanding, the number and percentage of students in vocational education remained fairly high (Table 4). However, these figures started to decline from the late 1990s onward, following the expansion of during the same period.

These movements are closely related to changes in the role of vocational education and training. In other words, education in the initial stage of industrialization focused more on equipping unskilled and undereducated youth with basic and job-specific skills, i.e., usually manual skills, to enter the labor market. However, as industrial structures became more sophisticated and educational opportunities broadened, social demands for vocational education and training gradually decreased. It instead assumed a bigger role among incumbent adult workers who needed to learn new skills to adapt to changing economic and technological environments. Thus in the 1990s, the vocational training system shifted from youth training to continuous training of adults.

Table 4: Upper Secondary Students by Type of Educational Program

	Total	Academic Students ('000)	Academic Share (%)	Vocational Students ('000)	Vocational Share (%)
1965	701	389	55.5	312	44.5
1970	889	408	45.9	481	54.1
1975	1,152	673	58.4	479	41.6
1980	1,353	748	55.3	605	44.7
1985	1,602	967	60.4	635	39.6
1990	1,683	1,096	65.1	587	34.9
1995	1,830	1,068	58.4	762	41.6
2000	1,975	1,193	60.4	764	38.7
2005	2,095	1,382	66.0	713	34.0
2010	2,253	1,561	69.3	692	30.7

Source: Adapted from Tables 3–8 in S. Paik. 2013. *2012 Modularization of Korea's Development Experience: Role of Private Schools in Korea's Educational Development*. Seoul: KDI School of Public Policy and Management.

Discussion

Vocational education and training in the Republic of Korea has changed and developed through various interactions with other fields within the entire education and training system. The rapid expansion of primary education and lower secondary education enabled adolescents to gain basic competencies for vocational education and training and reduced the supply of low-educated and unskilled labor. In rural areas, some youth still only finished middle school and had to find jobs in factories, but these gradually declined as enrollment rates in general education increased. This phenomenon became remarkable after the 1980s, when the so-called Lewis turning point—the end of an influx of low-skilled labor from rural to urban areas—appeared in the Republic of Korea (Sul 1992). As the supply of unskilled labor decreased, business owners dependent on unskilled labor gradually turned to production centered on machines and equipment, thus increasing demand for people with specialized vocational education and training. As such, the expansion of education below the secondary level had a positive effect on both supply and demand for TVET, and an overall negative effect on the expansion of higher education. First, as the overall structure of Korean industry became more knowledge-intensive, demand declined after the late 1980s until the early 1990s for vocational education and training centered on traditional manual skills. Furthermore, college graduates entering university after the expansion of higher education ended up in jobs that high school graduates had previously occupied, and demand for job seekers who had completed traditional vocational education and training diminished, since college graduates were willing to work for similar wages as high school graduates. It thus seems that expansion of higher education in general had a negative effect on traditional vocational education and training, which at that time was not properly equipped to adapt to changes in the labor market environment.

In addition, the increase in quantity of general education providers required more of the government budget, restricting expansion of investment to improve the quality of vocational education and training. In fact, complaints about inadequate quality had already existed in the 1960s and 1970s, when investment in these areas was emphasized. Yet, this phenomenon was inevitable given government budget constraints. It also became increasingly difficult to invest enormous resources in vocational education and training when the scale of general education continued to increase amid the country's high regard for education. Even after the 2000s, there was a period when investment in vocational education was intentionally expanded, such as during President Lee Myung-bak's administration (2008–2013). However, a significant increase in investment in vocational education was generally difficult amid overall pressure on the education budget from the expansion of general education.

A similar situation seems to be happening in other countries that are experiencing economic and social development. That is, the expansion of basic education will have a positive effect on the development of vocational education and training by equipping the workforce with basic yet needed competencies while reducing the supply of low-wage unskilled labor. However, if general education increases too rapidly, it could reduce the relative attractiveness of (and consequently, government support for) vocational education and training. In this regard, developing country governments should carefully monitor whether the entire education and training system is keeping pace with changes in the labor market as a whole, and adjust policies appropriately so that the education system and the labor markets can have a positive effect on each other.

Chapter 3
EXPERIENCE OF THE PEOPLE'S REPUBLIC OF CHINA

As with Chapter 2 on the Republic of Korea case study, this chapter on the People's Republic of China (PRC) will answer seven questions regarding (i) sector skills council, (ii) skills development for the future, (iii) technical and vocational education and training (TVET) policy in economic development, (iv) public–private partnership, (v) industry partnership, (vi) vocational qualifications framework, and (vii) pathways from general education to TVET. Answers to each of these questions will have three components. For each section, the chapter begins with introducing the country context, followed by a detailed discussion of the PRC's development experiences, then a summary of implications.

Given the limitations of policy diffusion and policy borrowing as reviewed in Chapter 1, this chapter is fully aware of the constraints in generalizing the experience of the PRC to other countries. The external validity of the PRC's solutions depends on its internal validity and the similarities across settings. The chapter aims to contextualize the PRC's responses to its skills challenges against its transition from a low-income country to middle-income country status. Two items in particular are noted here: the PRC's skills challenges, and its past experiences in borrowing technical and vocational education policies.

(i) Skills Challenges in the People's Republic of China

Economies in Asia and the Pacific feature diverse socioeconomic conditions and industry demands for skills and its societies at various development levels face significant challenges in matching the supply of skills with demand. Low-income countries have problems when graduates of vocational education institutions are unable to find employment, while managers of industrial and agricultural enterprises are short of skilled workers; families refrain from investing in their children's schooling unless they are confident of meaningful return (Easterly 2002).

Where existing institutions fail to bridge the gap between the supply of labor and the demand for it, poverty, unemployment, and underemployment in some sectors and regions can coexist with unmet labor demand in others for long periods of time. This condition was described by Arthur Lewis as a dual economy (Lewis 1954). Such a "skills market failure" is also characteristic of countries stuck in the "middle income trap," where countries can no longer compete in global product markets on the basis of low labor costs but are unable to compete in markets requiring highly skilled labor (Doner and Schneider 2016).

Skills market failure has also been evident in the PRC during its transition to middle-income status. Analysis of human capital accumulation and geographic distribution in the country suggests a looming human capital crisis in the near future (Khor et al. 2016, World Bank 2020). Recent studies demonstrate large regional variation and a huge urban–rural divide in spatial distribution of skilled labor, favoring eastern regions and metropolitan cities (Glaeser and Lu 2018). A shortage of skilled labor and the misallocation of human capital across sectors and regions have effectively slowed the PRC's growth, which is struggling to convert to an innovation-driven economy.

(ii) Past Experiences in Borrowing Technical and Vocational Education Policies

To solve skills market failure in developing economies, multinational organizations have proposed various policies for improving TVET as suggested in Chapter 1. But most promote best practices identified in advanced economies. These building blocks for successful TVET development are deeply embedded, however, in those countries' cultural, economic, and political environment. Without a proper institutional and cultural environment, many prior attempts at policy borrowing in TVET have failed terribly due to institutional mismatch, cultural mismatch, power differentials, and sociological mismatch (Lewis 2007).

Since the 1950s, the PRC has made several attempts to adopt TVET models from other countries. The most typical cases were the pursuit of the Soviet model in the 1950s and the piloting of the European models in the late 1990s. The partial success of both attempts can be attributed to a lack of the appropriate cultural, economic, and political environment in general; and the underdevelopment of a national skills formation system in particular—a result of historical legacy. As suggested by prior literature (Remington 2018), training practice transfers are difficult due to the institutional distance between countries in their political, economic, and educational institutions.

In the 1950s, the PRC's industrial base was extremely vulnerable. Unemployment rates were high especially among low-skilled workers. Historically, the PRC had no strong domestic tradition in TVET due to its late industrialization. Nearly 78% of unskilled labor had lost their jobs in Shanghai in 1952—21.5% for skilled labor, 0.87% for technicians, and 0.1% for engineers (Wang 2014, pp. 199–200). The geopolitical environment was also hostile to the PRC's development in the early 1950s. The outbreak of the Korean War put the PRC in direct confrontation with Western countries and cut off its access to foreign capital, technology, and human resources, except for assistance from the former Soviet Union. Accelerating industrialization and building a "complete" industrial system were major challenges for the new PRC at that time.

The country set out to pursue its Catch-Up Industrialization Strategy at the start of its First Five-Year Plan in 1953, following the Soviet example. This strategy emphasized prioritizing the development of heavy industries and implementing its "Planned Economy." The core mission of skills development then was to conduct mass literacy education and prepare skilled workers and technicians for key industrial projects. The PRC borrowed the former Soviet Union's workplace-based training model and established secondary vocational and technical schools affiliated to state-owned enterprises (SOEs) and imported Soviet-style technical grades for workers' evaluation in the 1950s.[21] Meanwhile, the government converted schools and universities to specialized TVET schools and colleges, tailoring to the skills needs of a narrowly defined heavy industry base. These reforms of vocational education and training at both the secondary and tertiary levels had far-reaching impact on skills accreditation and vocational qualification, matching skills supply with industry demand, TVET school and industry linkage, as well as relationships between general and vocational education.

Both systems were weakened during the Cultural Revolution from 1966–1976 (Risler 1989). The major limitation of the Soviet model was its overwhelming focus on upgrading skills of urban residents in SOEs, which left the majority of the urban and rural population as unskilled labor (Wang 2014). In the 1980s, the PRC's urban labor market suffered a concurrent lack of technicians and skilled workers and high levels of hidden unemployment and underemployment (World Bank 1992). The state started to search for new policy options for TVET sector reform, including feasible models abroad (Thøgersen 1990). The Education Commission intended to borrow Germany's dual system of apprenticeship training, while the Ministry of Labor was inclined to learn from the British National Vocational Qualifications system (Müller 2021). Both ministries formed their transnational actor coalitions with international partners. The Education Commission collaborated with the *Gesellschaft für Technische Zusammenarbeit*, a public organ of technical cooperation acting on behalf of the German government; and the Ministry of Labor cooperated with the British Embassy and the British Council. In addition, the World Bank played the role of facilitator through its aid and technical support projects. After 15 years of bureaucratic conflict between these two transnational actor coalitions, the attempt to transfer the dual system to the PRC failed (Stockmann et al. 2000). The British National Vocational Qualifications were partially adopted by the PRC government and integrated into a hybrid system in the late 1990s. Although the PRC introduced a system of vocational qualifications resembling the English system, the policy outcomes of Chinese National Vocational Qualifications left much to be desired due to the hybridization of National Vocational Qualifications with local institutions.[22]

Transferring social policy between different countries is notoriously difficult. For the PRC, both German and British models were relatively unique and thus problematic. Müller commented that "The German Dual System is based on historically grown institutions specific to Central Europe (Thelen 2004) and highly context dependent (Busemeyer 2015, Busemeyer and Iversen 2014), whereas NVQs as a policy innovation were based on a specifically British understanding of skills" (Brockmann, Clarke, and Winch 2011; 2021, p. 5). More generally, the dual system is embedded in the collective skill formation system in Germany (Busemeyer and Trampusch 2012b) and relied on various indirect and contingent effects (Wagner 1999). Prior studies have identified a series of challenges in transferring the dual system into other country contexts (Fortwengel 2017, Fortwengel and Jackson 2016), which were also evident in the PRC, the Russian Federation, and the US (Remington 2016, 2017, 2018).

---

[21] Interview with a former vice minister of education, 10 August 2020.
[22] Interviews with a former Communist Party of China (CPC) secretary and a former staff member of the Occupational Skills and Testing Authority Occupational Skills and Testing Authority, Ministry of Labor, 20 August 2020.

Despite several attempts at TVET policy borrowing, the PRC has been relatively successful in upgrading the skills of the majority of its population through compulsory elementary and secondary school education, and expanding vocational education and training at the upper secondary and tertiary level. During the past 40 years, human capital development strategy and economic development policies nationally have been clearly aligned. Meanwhile, regional or local experimentation in school–firm collaboration and involvement of local sector associations in skills formation have become defining characteristics of the PRC's TVET system.

The following sections briefly review the PRC's experiences in dealing with several challenging issues, such as identifying labor market skills needs, integrating TVET policy with economic development strategy, developing PPPs in the TVET sector, and bridging general and vocational education. The PRC's experiences during its transition to middle-income country status are potentially relevant for other economies in similar processes. On one hand, this chapter intends to assess whether it is possible or even beneficial to adopt international TVET best practices in Asian economies. On the other, given the similarities among the economies in Asia and the Pacific, the PRC's past experiences may be more relevant to its neighboring countries.[23]

3.1. Identification of Labor Market Skill Needs When Sector Skills Councils Could Not Be Formed

Background

In the early days of the PRC, the socioeconomic foundation of the country was extremely weak. The economy was on the brink of collapse. The total output value of national industry and agriculture was 46.6 billion yuan (CNY) in 1949, with more than 70% from the agriculture sector and 30% from industry. From 1949 to 1952, the average unemployment rate was 23.6% (Wang 2014). At the time, skills demand was hard to identify, because the overall skill level was too low to be measured. Manufacturing and agriculture were built on unskilled labor. Almost 80% of employed labor was illiterate and semi-illiterate. In 1949, there were only 564 vocational schools in the PRC, with a total enrollment of 77,095 students. Under such circumstances, relying on collective action of industry partners to identify skills demands was literally impossible.

Facing these socioeconomic challenges, the PRC government quickly converted to the planned economy and the "Catch-up Industrialization Strategy." With the limited role of employers, the skills formation system in the early days of the PRC was characterized by centralized workforce planning by the State Planning Commission and centralized implementation under the supervision of the State Economic Commission. The external labor market was abolished in the mid-1950s; instead, in urban areas, skills demand and supply was negotiated between line ministries and the State Planning Commission, and between the central and regional governments. The Ministry of

---

[23] Given that this chapter aims to illustrate the PRC's TVET sector development during its industrialization process and its transition to a middle-income country, the analysis mainly covers the country's experiences from 1949 to 2010, with few exceptions for recent policy development.

Labor played the role of skills coordinator under the guidance of the State Planning Commission and the State Economic Commission for SOEs in the urban sector, based on skills forecasting of the 5-year planning. This way of workforce planning was regarded as an "internal planning model" among line ministries.[24]

Progress

(i) Development of the External Planning Model

The legacy of the Planned Economy lasted nearly 40 years. In the 1980s, the "internal planning model" was widely criticized as the source of a concurrent lack of technical skills and high unemployment and underemployment in urban and rural areas. The external labor market was restored in the mid-1980s and the PRC decisively switched to the "Socialist Market Economy" system after the 14th Congress of the Communist Party of China in 1992. This transformation produced great impacts on the PRC's skills formation system. That regime was transformed into one characterized by open negotiation among the National Development Planning Commission (later the National Development and Reform Commission), Ministry of Labor, the Education Commission, and socialized sector organizations (former education or personnel departments of specialized economic ministries). This new model was considered an "external planning model."

The development of the "external planning model" was part of far-reaching labor market reforms in the PRC. The labor market gradually recovered in the 1980s. The ministries of labor and personnel implemented a "training first, employment afterwards" policy for SOE hiring and, in 1983, mandated that all new employees sign labor contracts. This policy was a sign of blue-collar labor market liberalization and put an end to both the employment quota system and lifetime employment. Under the new model, market forces played a significant role in balancing skills demand and supply.[25]

The development of the external planning model was facilitated by the introduction of national vocational qualifications system. During 1992, the PRC minister of labor inspected the Japanese Ministry of Labor's system of vocational qualifications. In 1993, the PRC Ministry of Labor then entered cooperation with the British Council and the Cultural and Education Section of the British Embassy. The new National Vocational Qualifications system was made concrete in 1994 when the Ministry of Labor and Ministry of Personnel issued regulations for a joint administrative framework for professional and vocational qualification certificates (Müller 2021). In 1995, the State Council issued its Interim Measures for Vocational Qualification Certificate System, endorsed by the Labor Law in 1995 and the Vocational Education Act in 1996. Under the new National Vocational Qualifications, certificates became mandatory for a list of 50 work types defined by the Ministry of Labor, and students from vocational schools and colleges increasingly had to acquire certificates to graduate, even in schools under the Education Commission.[26]

---

24   Interview with a former vice minister of education, 10 August 2020.
25   Footnote 25.
26   Interview with a TVET specialist and former staff of the Central Institute for Vocational and Technical Education, 25 August 2020.

(ii) Skill Planning with Sector Organization

Under the new NVQs, as former education or personnel departments of specialized economic ministries, the newly created socialized sector organizations played important roles in coordinating skills demand and supply.[27] On one hand, they had participated since the late 1990s in negotiations over classification of occupations and skills appraisal with the Ministry of Labor for the current and future labor force. On the other hand, as the labor security branches of their respective industries, these sector organizations organized their own industry guidance centers for skills appraisal and supervised the operation of industry-specific skills appraisal stations (see Box 1). In recent years, the Ministry of Human Resources and Social Security (a successor to the Ministry of Labor from 2008) had frequently collaborated with these sector organizations and their human resources branches to conduct surveys on current and future skills needs.[28]

Box 1: Sector Organization as an Industry Skills Engine

The China Nonferrous Metals Industry Human Resource Center positions itself as the industry skills engine, with its core mission to provide human resources and education services to members. It operates the China Nonferrous Metals Talent Network, the industry-wide recruitment and human resources service website.

The center is affiliated with the China Nonferrous Metals Corporation and State Bureau of Nonferrous Metal Industry and is part of the China Nonferrous Metals Industry Association.

The talent network generates first-hand information on distribution of skills by level and region through its labor market recruitment and skills appraisal services. It regularly collects information on industry skills accumulation from all 60 nonferrous metal industry-specific skills appraisal stations across the country. These stations offer appraisal for entry-level, semi-skilled, and skilled workers and final evaluation for technicians and advanced technicians. Moreover, the network can obtain current and future skills profiles of industry talent through its job search services for employers and vocational education institutions. Based on information provided by the network, the China Nonferrous Metals Industry Human Resource Center periodically is able to publish industry skills forecasting reports and make sector-wide training planning accordingly.[a]

---

[a] Interview with a technical and vocational education and training specialist at the China Nonferrous Metals Industry Human Resource Center, Beijing, 3 August 2020.

Source: Asian Development Bank.

---

[27] After the administrative consolidation reform of 1998, 10 specialized economic ministries were consolidated into other ministries or the State Council of the People's Republic of China. The education or personnel departments of these specialized economic ministries were transformed into trade associations or sector associations nationally. They receive budget allocation to fulfil certain administrative functions of their affiliated government agencies. In addition, they can generate revenues from training and other services for their member organizations. Their semi-government status was terminated after the General Office of the CPC Central Committee and the State Council jointly issued "The general plan of decoupling trade associations and chambers of commerce from administrative organs" in 2015 (State Council of the People's Republic of China 2015). The plan mandated that these socialized sector organizations to be independent from their affiliated government agencies and independent in terms of function, finance and assets, and personnel management.

[28] Interview with former director of Occupational Skill Testing Authority of Ministry of Human Resources and Social Security, vice president of Chinese Society for Technical and Vocational Education, 19 August 2020.

(iii) Regional Skills Partnership

Regional governments and local industry alliances are key players in soliciting local skill demands. In the manufacturing agglomerations in the Yangtze River and Pearl River deltas, prefectural city governments, county governments, and the Economic Development Zone Management Committee engage in skills development. When the national sector skills councils are weak, regional governments and their economic development branches can turn into the functional equivalent of sector skills councils and identify skills needs in the regional labor market (see Box 2). The active participation of local government often takes the form of PPPs, which creates semi-public organizations for matching skills demand and supply in industry agglomerate regions.

Box 2: Role of Local Government and Industry Alliances

The People's Republic of China and Singapore governments jointly established the Suzhou Industrial Park in 1994 and it has been number one for 4 consecutive years in the national ranking of economic development zones.

The industrial park government sponsors an intermediary agency, the Suzhou Industrial Park Human Resource Department, to steer local training provision. As the city government's extension agency, the department serves as the local human resources and social security bureau. It provides employment subsidies for firms, processes unemployment services, offers training for migrant workers, manages human resource archives for firms, and operates talent recruitment programs. It is also a commercial venture, recruiting employees on behalf of local employers, and gathering and publishing local labor market information, including annual reports on salary scales and labor shortage estimation for Suzhou Industrial Park's strategic industries such as electronic information manufacturing and machinery manufacturing.[a]

In recent years, Suzhou Industrial Park Human Resource Department began working with local industry alliances to identify current and future regional skills needs.

In Suzhou, many manufacturing enterprises participate in industry-specific alliances, such as for textile, home electronics, big data, nanotechnology, and intelligent manufacturing. For instance, with the sponsorship of the Ministry of Industry and Information Technology, the Suzhou Intelligent Manufacture Industry Alliance was established in 2017 and had more than 300 member organizations. The alliance regularly surveys skills needs among member organizations and shares information with the Suzhou Municipality Government and Federation of Industry and Commerce, the Bureau of Industry and Information Technology, the Human Resources and Social Security Bureau, and the Bureau of Statistics. Both Suzhou Industrial Park and Suzhou Municipality Government update and revise their human resources planning according to information provided by these local industry alliances and Suzhou Industrial Park Human Resource Department.[b]

[a] Interview with the Communist Party of China secretary of Suzhou Industrial Park Human Resource Department, Suzhou, 17 July 2017.
[b] Interview with a former president of vocational college in Suzhou Industrial Park, Beijing, 26 July 2020.
Source: Asian Development Bank.

Discussion

Comprehensive and accurate information on industry skills needs is a prerequisite for developing sensible skills and industry policy. In advanced economies, this task is often completed with the help of a well-functioning sector skills council. In less developed states, the industry base and employers are too weak to form capable sector skills councils. Alternatively, in the PRC and in many Asian countries, local governments share the responsibility of coordinating labor market supply and demand with socialized sector organizations and market human resources agencies.

Since the 1980s, the PRC government retreated from direct workforce planning to indirect guidance for its specialized economic ministries. Under the new external planning system, the labor ministry coordinated skills planning with support from other ministries and their regional branches. Nationally, the newly formed human resources branches of sector organizations are specialized in gathering industry demands and making industry-wide skills development plans. Regionally, local governments in industry agglomeration regions can develop intermediary organizations and work effectively with regional industry alliances to gauge current and future skills needs.

In the PRC's experience, the sector skills councils are neither a necessary nor sufficient condition for soliciting industry skills demand when the industry base is weak. The councils are part of the legacy of the trade guild in western countries dating back to the Middle Ages. Given late industrialization in most of Asia, these countries do not have trade guild traditions. In addition, civil society in these economies is not fully developed and the space for enterprises' collective action is quite limited. Under such circumstances, the local government is a possible yet less than perfect alternative for sector skills councils. The PRC case illustrates that socialized sector organizations and regional governments can turn into the functional equivalent of sector skills councils and identify skills needs in the labor market.

3.2. Economic Development through Aggressive Skills Policies Focusing on Future Skills Demand

Background

Policy on TVET can respond to current and future skills needs. The direction of national TVET policy is often determined by its skills formation system, a set of institutional arrangements including policies for risk sharing, skills standardization and accreditation, arrangements for training entities, and related socioeconomic institutions (Busemeyer and Trampusch 2012b).

Advanced economies have developed four types of TVET systems. Specifically, most countries with a collective skills formation system belong to a coordinated market economy, encouraging incremental innovation and emphasizing the importance of dual apprenticeship training to satisfy current industry needs. Thus, in countries such as Germany and Switzerland, TVET policy focuses more on satisfying the current skills needs of industry or firm. In contrast, economies with a statist skills formation system, characterized by high government and low firm involvement in pre-service training, are more open to both incremental and disruptive innovation (Hall and Soskice 2001). In turn, they are more likely to respond to current and future industry skills needs.

Under the planned economy, the PRC adopted many characteristics of the collective skills formation system and thus focused on responding to current industry or firm skill demand. From the early 1950s to the mid-1980s, the state and SOEs were heavily involved in pre-service training. Following the former Soviet Union's school-within-factory model, the PRC constructed numerous technical schools affiliated to SOEs. The key to the former Soviet Union's TVET model was teaching for production. Most technical schools were located in or adjunct to SOEs, and student trainees could obtain tacit knowledge and industrial know-how through close interaction with skilled workers on the shop floor (Chen 2008, 2009). Apprenticeship training was also critical for responding to industry skill demand. The training focused on firm- and industry-specific skills development with the assistance of training masters (Wang 2014). The state adopted the former Soviet Union's pedagogical documents, textbooks, examinations, and evaluation standards, and invited Russian experts as consultants for teacher training programs and curriculum development (Fang, Liu, and Fu 2009).

Progress

After 30 years of experimentation with the former Soviet Union's workplace-based training model, the central authority dramatically switched to a school-based training system in the 1990s. As a decisive step toward the Socialist Market Economy, the State Council consolidated 10 of its specialized economic ministries into the National Development Planning Commission in 1998, and transferred the line ministries' affiliated secondary and tertiary education institutions to the Ministry of Education or regional governments. Most SOE-affiliated vocational institutions were transferred to local education authorities and later converted to a school-based training system.

Because secondary and tertiary vocational education were no longer sponsored and managed by the Ministry of Labor or specialized economic ministries such as the Ministry of Machinery Industry and the Ministry of Electronic Industry, their curriculums and pedagogical practices soon became outdated. To remain responsive to current and future industry needs, the Ministry of Education strongly encouraged reintegration of VET schools and enterprises through multiple channels, including piloting modern apprenticeship programs, promoting PPPs, establishing vocational education groups with industry partners, launching vocational education parks, and introducing mandatory skills appraisal and certification of vocational graduates.

One of the most promising policy instruments is the creation of an industry teaching and education steering committee for strategic industries, which establishes buffer institutions to connect VET institutions and enterprises in skills formation. To bring industry back to school-based training, the Ministry of Education has sponsored 62 industry teaching and education steering committees for TVET since 1999. The steering committees consist of experts from TVET institutions, academia, industry associations, and enterprises.

The core objectives of the committees include analyzing industry and occupational skill demands; proposing competence, knowledge, and skills requirements for industry professionals and skilled workers; recommending curriculum and pedagogy development in TVET institutions; participating in construction of standardized teaching systems for VET; promoting the integration of vocational qualification certificate and academic teaching; and facilitating school–firm collaboration, etc.

Among the steering committees, the National Petroleum and Chemical Vocational Education and Teaching Steering Committee had successfully connected industry skill demands with school-based training (see Box 3).

Box 3: Bridging Industry Needs with Vocational Education and Training

Skill shortages and mismatches are a serious challenge for the People's Republic of China (PRC) petroleum and chemical industry. The industry employs more than 20 million workers, but technical and vocational education and training (TVET) institutions only provide 0.2 million to 0.3 million graduates each year. Due to the limited skills supply, about 20% of the industry's current workforce has no prior training or appropriate degrees for their jobs.

To improve its talent pipeline, China Education Association of Chemical Industry was established in 1995. The association works closely with the National Bureau of Statistics and Chemical Data Center of China Petroleum and Chemical Industry Association to collect information on structure, size, and education and training distribution of the current labor force. The association has published industry skills forecasting reports every other year since 2007 and organized national surveys of enterprises' skill demands in 2014 and 2017.

Since 2011, the China Education Association of Chemical Industry has hosted the National Petroleum and Chemical Vocational Education and Teaching Steering Committee. More than 40% of its 700 committee members are industry experts. The committee provides industry experts opportunities to participate in teaching and curriculum standard revision of VET institutions, textbook development, faculty professional development, and practical training-based standards development. By bringing industry experts closer to the TVET education process, National Petroleum and Chemical Vocational Education and Teaching Steering Committee serves as a bridge linking skill demand with supply.[a]

[a] Interview with a vice president of the China Education Association of Chemical Industry, Beijing, 22 August 2020.
Source: Asian Development Bank.

As buffer organizations, the industry teaching and education steering committees also lead promotion of future skills development. For instance, the National Machinery Vocational Education and Teaching Steering Committee played important roles in the Twelfth Five-Year Plan for Machinery Industry and the Thirteenth Five-Year Plan for Machinery Industry by editing industry-wide skills talent cultivating plans. To support the implementation of the ambitious "Made in China 2020," this committee conducted detailed human resource analysis for new and strategic industries, including industry robots and new energy vehicles. Based on these forecasts, the committee launched several key college specialty construction projects, to guide specialty development in vocational colleges. In summary, these industry teaching and education steering committees became effective intermediaries between the Ministry of Education, TVET institutions, sector organizations, and industry partners.

Discussion

The PRC experience implies that a nation's skills development policies are deeply embedded in its industrialization strategy and economic development stage. Market reforms in the early 1980s deregulated the PRC labor market and spared the training function of SOEs. As a result, the PRC's training system became more similar to the statist skills formation system of the Nordic countries, where the state provided full support for school-based training and firms' involvement was marginal. As the national coordinator of school-based training, the Ministry of Education had to encourage VET school–firm collaboration to respond to industry demands.

Following the examples of the statist skills formation system, the PRC created new types of intermediary organizations to facilitate the cooperation of VET schools and enterprises. The Ministry of Education and human resources branches of sector organizations' co-established industry teaching and education steering committees as buffer organizations. Some buffer organizations were successfully bringing enterprises back to school-based training and making TVET policy relevant for current and future skills needs. They set an organizational environment conducive for industry experts' participation in curriculum and practical training development in TVET institutions and for integrating school leaders with industry talent planning.

In conclusion, these buffer organizations increased the quality and relevance of school-based training, which are responding to current and future industry demand. Although enterprises and VET institutions were separated after the PRC's shift toward the statist skills formation system, they can be formally or informally connected by these newly established intermediaries.

3.3. Status of Technical and Vocational Education and Training Policy within the Whole Economic Development Strategy and the Conditions for Successful Integration

Background

The PRC's skills development policy making has always been in line with the nation's key strategy and policy priorities. Workforce planning has been an integral part of the PRC's planned economy since the early 1950s. The Catch-Up Industrialization Strategy at that time was implemented through the construction of 694 key industrial projects. Following the Soviet model, the central government established new technical schools for almost all key projects to satisfy their needs for skilled workers and entry-level technicians. After the launch of the Reform and Open-Up Policy, the PRC's TVET policies were centered on responding to the needs of economic development under the Socialist Market Economy.

What is the policy process of integrating TVET as an essential part of national development strategy in the PRC? Who are the key policy agents in the policy making process? How are key policies formed? What are the characteristics of the policy process? Prior studies analyzed public policy making in the PRC through the elitist model, the factionalism model, and the bureaucratic organization model (Chen 2012). The bureaucratic organization model highlights that frequent interactions between high-level elites and the bureaucracy lead to the making of public policy. Once high-level elites make their decisions, national bureaucracies take on the responsibility of policy making and implementation, such as feasibility research, consensus coordination, funding allocation, policy content clarification, and further implementation (Lieberthal and Oksenberg 1998).

The early development of the PRC's TVET policy closely followed the logic of the bureaucratic organization model in its political mobilization, national strategy formulation, and sector implementation. In the 1950s, political mobilization built a tight link between skills formation and the national imperative of accelerated industrialization. To satisfy the skills needs of accelerated industrialization, the PRC's constitution and Communist Party of China's national congress highlighted the importance of technical education at secondary level (Fang, Liu, and Fu 2009). The political will of the CPC leaders was soon transformed into a national TVET development strategy. The Government Administration Council announced its Decision on the Reform of Education System (1951) and prioritized adjusting secondary education structure and integrating technical education as part of secondary education. Sector authorities swiftly adopted this national strategy into their specific sector planning and TVET policies. The Ministry of Education organized two national secondary vocational education meetings in 1951 and 1956 and highlighted the importance of rapidly expanding school-based secondary TVET. The Ministry of Labor encouraged specialized economic ministries to create and operate their own technical schools (Fang, Liu, and Fu 2009). The state commanded the heights in education and training sector and intentionally integrated TVET policy with its accelerated industrialization policy.

Process

After the Cultural Revolution, workplace-based training and secondary vocational and technical education were seriously damaged. Almost all secondary technical schools were closed in the early 1970s: many schools were turned into factories, teaching facilities and libraries were destroyed, and students were dismissed. To revive the science and education system in the PRC, then Chairman Deng Xiaoping announced that "Science and technology are the first productive force" in 1978. The marketization reform became the main driver for the political mobilization for integrating TVET policy with national economic strategy. In 1984, the Decision of the CPC Central Committee on Reform of Economic System marked the beginning of marketization reform. The 14th National Congress of the CPC in 1992 claimed that the PRC would adopt the Socialist Market Economy. The subsequent labor market reforms had profound impacts on the education and training system (Figure 4).

The successful political mobilization by the top leadership facilitated the emergence of a new national strategy for vocational education. In 1985, the CPC Standing Committee issued its *Decision on Reform of Education System*, which reiterated the importance of adjusting the education system to meet the multifaceted needs of social and economic development. The State Council released its *Outline for Educational Reform and Development* in 1993, which foreshadowed changes in the PRC's skills formation system, from a predominantly workplace-based training system to a school-based training system.

Responding to the new national TVET strategy, the Ministry of Education and Ministry of Labor quickly announced their respective sector strategies. The Ministry of Education organized three consecutive national vocational education meetings in 1986, 1991, and 1996, with the intent to enlarge the school-based TVET sector. Meanwhile, the Ministry of Labor proposed to establish a new technical school model under the Socialist Market Economy, according to its Opinion on Deepening Reform of Technical Schools (1989). This opinion terminated the compulsory

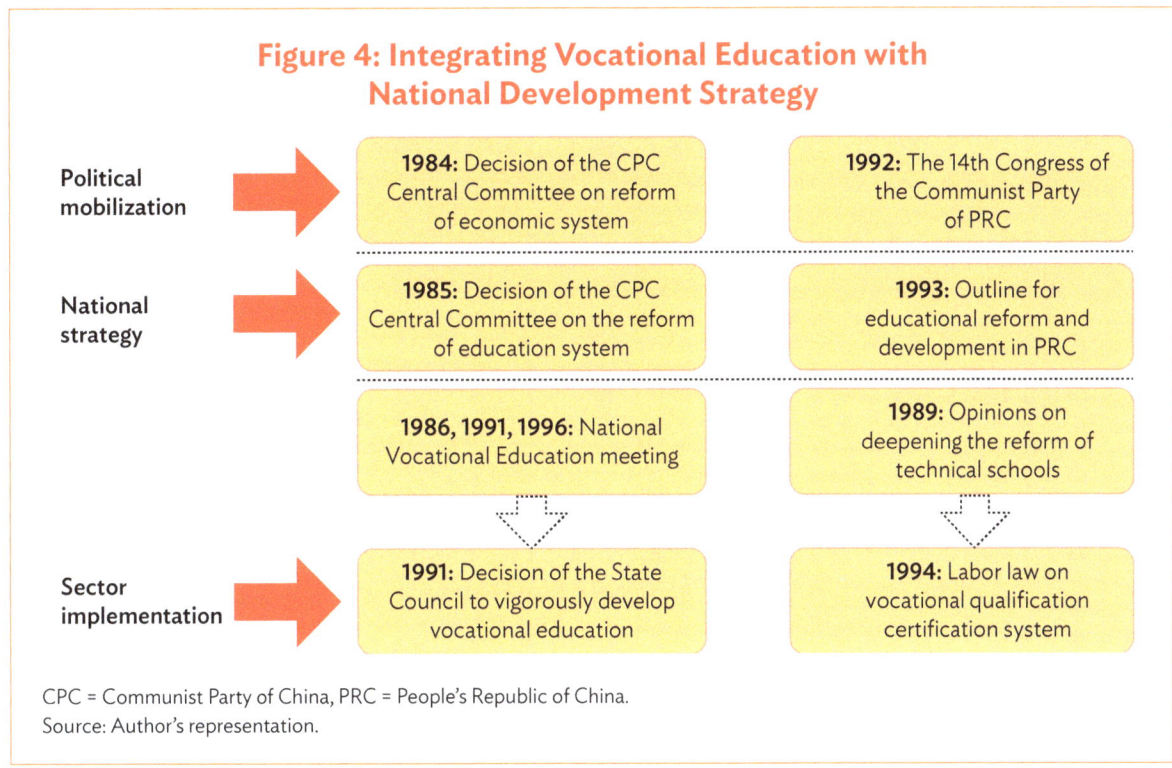

Figure 4: Integrating Vocational Education with National Development Strategy

CPC = Communist Party of China, PRC = People's Republic of China.
Source: Author's representation.

graduate employment allocation system for VET school graduates and proposed to implement a double-certificate system. This new system mandated that all VET graduates should have both academic degrees and technical grade certificates. The Labor Law of 1994 introduced the new NVQs and integrated TVET with the new labor market institutions.[29]

Discussion

The PRC's experiences show a tight coupling between its TVET policy and economic development strategy at the national level, a defining characteristic of the development state approach (Ashton et al. 2002; Green et al. 1999; Lauder, Brown, and Ashton 2017).

As in advanced economies, TVET was not an independent policy area in the PRC. It was part of its skills formation system and was deeply embedded in its national development strategy and sector implementation planning. Based on a review of historical documents and interviews with policy makers, it is clear that the PRC's key policy makers integrated TVET policy with national development strategy in three stages.

(i) *Political mobilization by high-level authorities.* In the PRC context, key decision-making agencies consist of the Political Bureau of the Central Committee, the Central Military Commission, and the Executive Meeting of the State Council. These high-level elites held instrumental views on vocational education and training.

---

[29] Interview with a former vice minister of education, 10 August 2020.

(i) Following the former Soviet Union's experience, the leadership core considered TVET an integral part of workforce planning under the planned economy. After the transition to the Socialist Market Economy, the central authority re-evaluated its preferences regarding TVET and prioritized development of TVET over general education at the secondary level. Endorsement from high-level authorities provided legitimacy for integrating TVET policy with the national development strategy.

(ii) *Transforming the leadership core's political will into national strategy.* To implement the political will of high-level authorities, the State Council and specialized taskforces were able to quickly translate the political ideology into a national development strategy, including 5-year planning and sector-specific reform guidelines or decisions.

(iii) *Policy making and implementation by sector authorities.* After the national development strategy clarified the specific workforce planning goals, sector authorities—including education and labor ministries and specialized line ministries—made their own sector planning in a timely manner. Given that the key characteristic of the bureaucratic organization model is "being responsible to the superior," the sector authorities were responsible for integrating TVET policy with the national development strategy under the supervision of the State Council.

Although it is often the responsibility of strong economic ministries to integrate TVET with the national development strategy (Ashton et al. 2002; Green et al. 1999; Lauder, Brown, and Ashton 2017), the PRC's case demonstrates the importance of political mobilization from the very top. Without the clear endorsement of the leadership core, the strong economic ministries were less likely to take action. Because the PRC is a regionally decentralized system (Xu 2011), political mobilization is critical for diffusing the new state ideology and mobilizing resources for implementing the new development strategy.

The strong role of political mobilization in the PRC's case may not be extrapolated to other economies with different political regimes. The three-stage model—political mobilization, making of national strategy, and sector implementation—developed from the PRC experience is simply one option for integrating skills policy with national economic strategy. Other countries should consider their own ways of policy integration according to their country contexts.

3.4. Pursuit of Public–Private Partnerships and the Role of Government in Driving Such Cooperation

Background

Despite rapid growth of the VET sector since the late 1990s, providing industry with relevant skills remains a key challenge in the PRC. The challenge originates from the failed collective action in the training market. On one hand, interfirm collaboration in training was rare due to the risk of poaching in a highly flexible labor market. Consequently, firms were reluctant to invest in training and hire workers from the external labor market. An enterprise's nontraining strategy leads to insufficient skills supply. On the other hand, school–firm training collaboration is hard to achieve. Working intensively with firms implies high costs and uncertain benefits for VET schools. They have to align their curriculum with industry needs, ensure their training meets sector standards, and provide professional training and new facilities. Collaboration implies high costs for firms, including investment in time, funding, and human resources. In addition to the risk of poaching, firms are worried about training quality and leakage of vital production technology (Marques 2017).

Public–private partnerships are a promising policy solution to failures in the skills market.[30] Since the early 2000s, the PRC's TVET schools have adopted various PPP-type collaborations, including PPPs for education services, facility construction, and overall operation (Han 2016). Political support for the TVET PPP in the leadership core has been growing over time. The State Council called for deepening the cooperation between vocational schools and industry partners in its Opinion on Accelerating the Development of Vocational Education (2005). In addition, it proposed to establish new forms of private vocational schools through mixed ownership in 2014. The Private Education Promotion Act, 2002 and its 2016 amendment legitimized PPPs as one kind of social investment for education (Yang 2017).

Under this general national framework for promoting PPPs, the Ministry of Finance and the National Development and Reform Committee have been actively engaged in setting the policy framework for vocational PPP programs. As strong budgeting and planning authorities, they consider PPPs a viable way for strengthening school–firm partnership and aligning TVET policy with national development strategy. The state council issued the Measures to Promoting School Enterprise Cooperation in Vocational Schools in 2017. It provided regulations on the organization form, the eligibility condition, the type of cooperation, rights and responsibilities of all parties, the content of cooperation agreement, and procedural management for school-enterprise cooperation. The National Development and Reform Committee plus five other ministries jointly issued the Implementation Measures for the Construction of Enterprises Integrating Production and Education in 2019, which stipulated that qualified enterprises should enjoy preferential policies in project approval, purchase services, tax exemption and tax cuts, financial support, and land use priority.

Progress

The key challenge facing the PRC's skills formation system is to bring industries back to joint skills investment. PPP is a result rather than a solution to the "collective action dilemma"[31] in the training market. In the PRC's context, PPPs in the vocational sector are organized and facilitated by intermediary organizations. They connect employers, education and training institutions, government bodies, and other organizations. Many kinds of organizations can serve as intermediaries, including sector and regional business chambers, industry councils, educational organizations and training centers, and government agencies (Remington and Yang 2020).

---

[30] Per the ADB's *Public-Private Partnership Handbook,* PPPs present a framework that—while engaging the private sector—acknowledges and structures the role for government in ensuring that social obligations are met and successful sector reforms and public investments achieved (ADB 2008, p. 1).

[31] The collective action dilemma refers to the skills market failure, such as insufficient supply or demand for skills. The government, employers, and employees cannot reach agreement on skills investment.

The PRC's regions have developed various forms of intermediaries. Depending on the degree of interfirm cooperation and school–firm collaboration, there are four PPP cooperation models (Figure 5), including loose coupling model; firm-led model; government-led model; and joint-led model. Each cooperation model differs in terms of identity of intermediary and participants, type of training cooperation, way of cost sharing, skills formation location, and type of skills produced (Yang 2017).

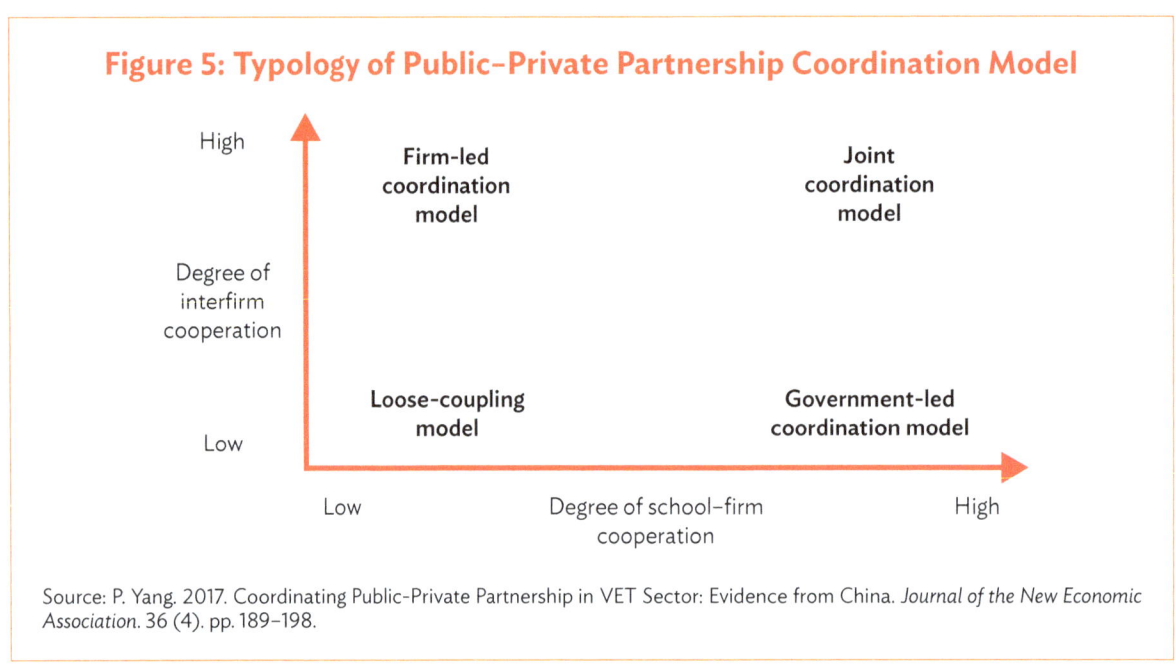

Figure 5: Typology of Public–Private Partnership Coordination Model

Source: P. Yang. 2017. Coordinating Public-Private Partnership in VET Sector: Evidence from China. *Journal of the New Economic Association*. 36 (4). pp. 189–198.

The bottom right corner of Figure 5 is the government-led model, in which school–firm cooperation is strong and interfirm collaboration is weak. Local governments often designate or establish a skills intermediary. This model is quite common in regions with a high concentration of manufacture industries, where industry agglomeration leads to high demands for industry-specific skills. Typically, regional economic zone or industrial park authorities can supervise and monitor local skills production, with support from industrial and educational partners (see Box 4). In this case, governments are major investors for school-based training in TVET institutions, local public training bases, and interfirm training centers.

The firm-led model in Figure 5 is characterized by high-level interfirm collaboration and low school–firm cooperation. In industrial agglomeration regions, dominant firms are leaders in training partnership, and business associations can play the role of intermediary (see Box 5). Firms and governments share the costs of training programs for industry-specific skills, including school-based training, intensive workplace training, and dual training in firms.

Box 4: Government Agency as Intermediary Agency

A strong intermediary organization exists in training public–private partnership coordination in Suzhou Industrial Park, Jiangsu province. Suzhou Industrial Park Government sponsored an intermediary agency, Suzhou Industrial Park Human Resource Department, to steer local training provision. It had nurtured strategic partnership between firms and the vocational colleges to produce industry-specific or firm-specific skills in Suzhou Industrial Park.

The department worked closely with Suzhou Industrial Park's affiliated Institute of Vocational Technology. To ensure the institute's training could meet local industry qualification standards, the human resource department served as the institute's board of trustees member. It regularly participated in the institute's curriculum development and jointly sponsored in campus, interfirm public training centers for small and medium-sized enterprises. Furthermore, it supported the institute's Samsung Semiconductor Engineering Academy, offering baccalaureate and master's degrees for Samsung and its supply chain partners.

The Suzhou Industrial Park Human Resource Department helped Bosch Auto Parts Company in Suzhou launch its dual apprenticeship training program with three local vocational colleges. This intermediary has enhanced the local training supply through its interfirm training programs by coordinating interests of firms of various sizes in similar industries.[a]

[a] Interview with a Communist Party of China secretary of Suzhou Industrial Park Human Resource Department, Suzhou, 17 July 2017; Interview with a former president of vocational college president in Suzhou Industrial Park, Beijing, 26 July 2020.

Source: Asian Development Bank.

Box 5: Business Associations as Public–Private Partnership Intermediaries

Business associations can turn into a catalyst in local training markets.

Since the early 1990s, German manufacturing firms started to cluster in Taicang City, Jiangsu Province. In Taicang's Sino-Germany SME Cooperation Demonstration Zone. In 2000, the business association of Germany's small and medium-sized enterprises in Taicang negotiated with Taicang Vocational Secondary School to establish an interfirm training center following the dual apprentice training model. The association worked as an intermediary and convinced local German firms not to poach from each other and to make joint skills investment.

Guided by the business association and the German Chamber of Commerce in China (AHK-Shanghai), this school created the Taicang Professional Workers Training Center in 2001. Following the success of the center, AHK-Shanghai and Suzhou Chien-Shiung Institute of Technology jointly founded the AHK Professional Workers Training Center in 2007.

As an intermediary, the German business association also encouraged dominant firms to create their own dual training programs with vocational education and training institutions. Schlaeffler Greater China collaborated with Taicang Vocational Secondary School and established the Schaeffler Training Center in 2005; and Häring China launched its Häring Academy with this school in 2013.

The Suzhou Chien-Shiung Institute of Technology, Tongji University, and Schlaeffler Greater China jointly initiated a dual baccalaureate program with support from the German business association and AHK-Shanghai in 2015.[a]

[a] P. Yang. 2017. Coordinating Public–Private Partnership in VET Sector: Evidence from China. *Journal of the New Economic Association*. 36 (4). pp. 189–198.

Source: Asian Development Bank.

The joint-led model is one in which interfirm cooperation and school–firm collaboration are both strong. Government agencies and industry associations together can coordinate local skills production, with large private firms and governments as the main investors while vocational colleges, research universities, industry organizations, and SMEs participate in skills production.

The model is conducive to the production of general skills and industry-specific skills. A typical example occurred in the sanitary product and shoes and textile industries in Jinjiang City, Fujian Province. With the support of Jinjiang Bureau of Science and Technology and Bureau of Education, Heng'an International Group collaborated with Fujian Business Association for Sanitary Products and Quanzhou Institute of Technology and offered an associate degree program specialized in sanitary machinery manufacture and maintenance. This specialized school responded to skills needs in upstream and downstream firms in that industry. Under the joint-led model, the Bureau of Education enforced training regulations and ensured vocational colleges met industry qualification standards, and the business association reconciled the training needs of different firms (Remington and Yang 2020, Yang 2017).

Discussion

As Chapter 1 illustrated, the most common reason for supporting a PPP is its ability to increase industry involvement. As such, the PRC promoted school–firm collaborations in PPPs to address the collective action dilemma. In theory, the PPP can solve this problem by bringing cooperation incentives for social partners. Joint investment by social partners and government can ensure each partner's action is consistent with its commitments (Yang 2017).

In the PRC's VET sector, there are multiple examples of the joint coordination model, the firm-led model, and the government-led model for PPP cooperation. Notably, sector and regional business associations, industry councils, educational organizations and training centers, and government agencies can function as an intermediary in PPP collaboration. Particularly, according to the PRC experience, government can play multiple roles in vocational PPPs.

(i) *Government as environment.* Local government can set up a proper policy framework for vocational PPP. In the PRC context, VET schools and firms make skills investment decisions according to their own needs and policy incentives. This is the typical case of the firm-led model where enterprises in a monopoly can form strategic skills partnerships with local VET institutions and SMEs in the same industry, responding to policy incentives in the local skills market.

(ii) *Government as intermediary.* The government can play the role of intermediary between VET schools and firms, and facilitate their cooperation in vocational education and training. This is observed in both government-led and joint-coordination model, where government agencies mobilize resources in industry, firms, and VET schools to create an "industry skill commons" for the local market.

What were the conditions under which PPP incentives and regulations could work effectively? One key observation from the PRC is that an intermediary is crucial for setting up effective incentives and regulations. An intermediary is an institutional solution to the collective action dilemma, particularly effective in solving the commitment problem of organizational partners. From the perspective of institutional capacities, the effective intermediary can coordinate the activities of government, firms, and schools; ensure their commitments; and generate credibility in rewards or sanctions (Doner and Schneider 2016). Without proper intermediaries, skills partnership can be fragile and unsustainable.

Intermediaries are critical for ensuring that organizational actions are consistent with commitments. They can monitor each partner's activity, including developing training standards and qualifications and participating in curriculum development and examination. To ensure partners' commitments are credible, intermediaries can enforce regulations on responsibilities of firms, schools, and individuals. Therefore, intermediaries are able to enhance interfirm cooperation and school–firm collaboration by reinforcing credible commitment.

3.5. Policies and Measures for Promoting Employer-Led Training

Background

Rapid technology innovation has had a huge impact on the relation between industry and the education sector. New technology ties industries to universities, given that research-intensive universities are capable of producing basic and applied research and transferring new technology to industry (Nelson 1994). The implementation of new manufacturing technologies can tie industry to vocational education institutions, given VET's critical role in regional innovation systems (Lund and Karlsen 2020).

A regional innovation system is a framework for close interfirm interactions, knowledge sharing, policy support infrastructure, and a sociocultural and institutional environment. The broad definition of a regional innovation system includes "all parts and aspects of the economic structure and the institutional setup affecting learning" (Lundvall 2010, p. 13). The regional innovation system corresponds to the "doing, using and interaction mode of innovation," where innovative activity is based on a synthetic knowledge base, i.e., experiences and competencies from practice gained through on-the-job training and education, rather than based on research activities (Asheim 2012). Vocational education institutions are specialized in producing synthetic knowledge through school-based practical training. However, keeping TVET programs in line with rapidly changing technologies is a challenge facing the PRC's VET institutions.

Progress

(i) National Policy Framework and Typology of Skills Partnership

The PRC government had endorsed the principle of school–firm collaboration since the 1990s, and was first introduced as a national TVET strategy in 1993. The Vocational Education Act of 1996 reinforced that VET schools should work closely with industry partners in training and skills certification. In March 2014, Prime Minister Li Keqiang called for the development of a modern, employment-oriented vocational education system. These policies were a prelude to the State Council's 2014 decision on developing a "contemporary apprenticeship system," a hybrid version of the European dual training system with elements of the PRC's domestic institutions. The policy manifested itself in many forms. For instance, the Ministry of Education launched its Modern Apprenticeship Pilot Program in 2014, the National Development and Reform Committee initiated its dual apprenticeship training pilot program in 2015, and the Ministry of Human Resources and Social Security established its Enterprise New Apprenticeship Pilot in 2018.

In the wake of the marginalization of enterprises in skills training, the leadership core reassessed its preference for school-based and publicly sponsored training. Following Germany's dual training model, the central authority in 2010 claimed enterprises would be on equal footing in skills partnerships with VET schools. In this circumstance, PPPs became the preferred model for school–firm collaboration. The 19th National CPC Party Congress in 2017 explicitly called for deepening such a collaboration, and the State Council immediately issued an opinion on promoting education–industry integration. In early 2018, a broad policy coalition was formed around this initiative with active participation of ministries, including the Ministry of Education, National Development and Reform Committee, Ministry of Industry and Information Technology, Ministry of Finance, Ministry of Human Resources and Social Security, and the State Taxation Administration. As a further step toward promoting skills partnerships, the Implementation Plan of National Vocational Education Reform (2019) intended to transform the predominantly government-run TVET model into a PPP-type skills partnership.

In practice, regional governments and enterprises have nurtured various local skills partnerships in line with the national strategy. One simple way to classify skills partnerships among the PRC firms, VET schools, and local authority is to divide partnerships along two dimensions: the degree of coordination across firms operating in the same labor market, and the costliness of resources committed to cooperative arrangements on the part of partners (Remington and Yang 2020). For simplicity's sake, the first dimension is defined as "breadth" and the second as "depth." These two dimensions yield a two-by-two matrix (Figure 6).

In the top right cell is the solidaristic model with a high collaboration among partners and high investment by partners. Characteristics of such systems are national-level PPPs among the state, schools, employers, and trade unions. The liberal model is characterized by low coordination among firms and low cooperation between firms and government. Typically, individuals pay for their

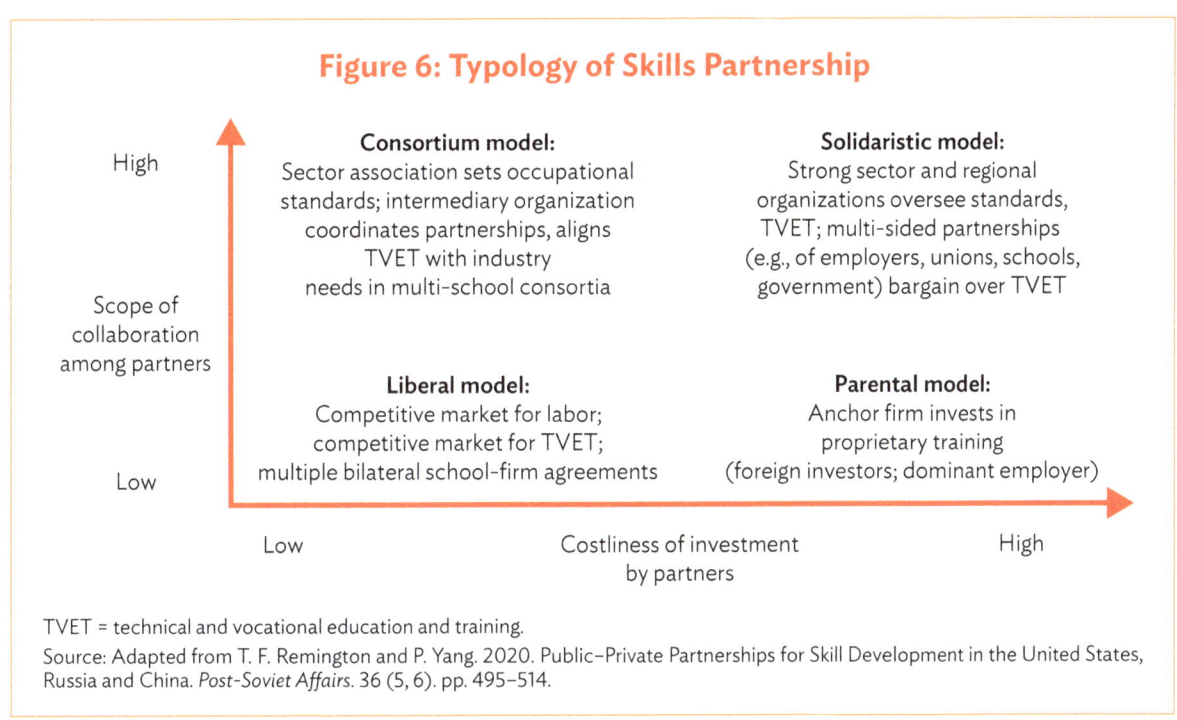

Figure 6: Typology of Skills Partnership

TVET = technical and vocational education and training.
Source: Adapted from T. F. Remington and P. Yang. 2020. Public–Private Partnerships for Skill Development in the United States, Russia and China. *Post-Soviet Affairs*. 36 (5, 6). pp. 495–514.

own education and training beyond K-12 general education. Bilateral school–firm collaboration exists, but interfirm cooperation is rare. The consortium model consists of multiple partners, but investments made by partners are limited. Multiple firms collaborate on training with one or several educational institutions, aligning their common interests with industry-specific skills. In the bottom right quadrant is the "parental" model, with few partners, but heavy investment from each partner. With the encouragement of local government, local education authorities work with the dominant firms ensuring that education and training not only provide firm-specific skills, but also generic skills and knowledge transferable to other firms.

(ii) Development of Regional Skills Partnership

The policy summary indicates that deepening school–firm collaboration and education–industry integration are considered key strategies for promoting school–industry linkage in the PRC. Examples exist of parental models and consortia models in various regions of the PRC (Remington and Yang 2020). Parental arrangements are often found in strategic industries such as railroads and the aeronautical industry (see Box 6). Under this model, industry associations and dominant firms set occupational and educational standards for technical schools and align the curriculum with industry needs.

> **Box 6: Parental Model in Strategic Industry**
>
> The China Railway Rolling Stock Corporation, the largest manufacturer of rolling stock in the world—based in Hunan Province but operating in over 100 countries—relies closely on the Hunan Vocational College of Railway Technology and the Hunan Railway Professional Technology College for training, assessment, and certification of trainees.[a]
>
> Both colleges provide apprentice training programs for the China Railway Rolling Stock Corporation, tailoring to its skills needs in domestic and international markets. The corporation shapes the content of training programs by providing experienced trainers, donating facilities for practical training, and offering attractive entry-level positions for college graduates. Other enterprises can use these colleges for in-house training under this parental model.
>
> Both colleges provide examinations for entry- and advanced-level technicians and engineers on behalf of railway enterprises and the China Railway Rolling Stock Corporation. The colleges grant industry-wide recognizable certificates for trainees who pass their examinations. They also coordinate training for metropolitan rail transportation companies in Changsha City.
>
> ---
> [a] Interviews with a vice president of Hunan Vocational College of Railway Technology and a dean for industry collaboration of Hunan Railway Professional Technology College, Zhuzhou, 27 June 2017.
> Source: Asian Development Bank.

Partnerships of the consortium type also exist in the PRC consumer industry. In Guangdong Province, specialized towns work with local vocational colleges to provide industry-wide training for local firms, sharing costs among colleges, firms, and individuals. In Guzhen of Zhongshan City, the Zhongshan Polytechnic opened a specialized school for the local lighting industry, supported by the township government and the lighting industry association. The township government invested more than CNY21 million in land, building, and facilities, and the polytechnic invested CNY8.5 million for teaching equipment, faculty, and student services. The school provided on-the-job training for lighting art designers, production line managers, and other professionals in local small and medium-sized enterprises.

In recent years, disruptive technologies and Industry 4.0 have dramatically changed industry skill demands and skills partnerships between VET institutions and firms. Given the unlikely prediction that technologically advanced companies can satisfy their skills needs only with graduates from research universities, they would need vocational college graduates when they implement new manufacturing technologies (Lund and Karlsen 2020). Huawei's collaboration with Shenzhen Polytechnic is a good example of dynamic skills partnership in an emerging industry (see Box 7).

Box 7: Shenzhen Polytechnic Co-Evolving with Huawei

Huawei Technologies Co. Ltd is a leading information and communication technology (ICT) solutions provider. It has established end-to-end advantages in telecom networks, devices, and cloud computing. Its products and solutions have been deployed in over 161 countries.

As a dominant firm in the information technology industry of the People's Republic of China, Huawei has been using its market position to prepare its talent pipeline through research universities and vocational colleges. In its collaboration with Shenzhen Polytechnic, Huawei went beyond the traditional parental model and adopted a more flexible and dynamic approach.

Over the years, Shenzhen Polytechnic has created a strategic partnership with Huawei. The polytechnic's College of Electronics and Communication Engineering has gradually adopted Huawei's certification system and integrated certificate training with its curriculum. Specifically, Huawei helped the college divide the certificate training into multiple stages, multiple layers, and multiple levels, so that students could attend a series of Huawei certification courses and receive related certifications.

As the enterprise certification standards continued to upgrade, so did the certification-based courses. The curriculum now covered a wide range of customized courses such as Huawei internet protocol data communication technology, mobile communication technology, and Huawei cloud computing, cloud service, and network. Most graduates are referred to as Huawei Certificated ICT Associate or Huawei Certified ICT Professional.

By the end of 2018, 79 outstanding students had passed the examination for Huawei Certified ICT Expert, the highest level in Huawei's certification system. The new curriculum with its distinctive industry characteristics was well received by other firms in the same industry.

This dynamic skills partnership has become a part of the Huawei ICT Academy, comprised of 927 colleges and universities in 72 countries as of 2019. Its curriculum development standards have been diffused to other Belt-and-Road countries.[a]

In addition, 11 of the Huawei authorized online courses developed by Shenzhen Polytechnic have been used for on-the-job training for more than 6.86 million Chinese engineers. In 2018, Shenzhen Polytechnic won the National Teaching Achievement Award. Through this active partnership, state-of-the-art industry technology can quickly be disseminated to vocational education and training institutions, and these institutions can codify the knowledge into curriculums that can be shared with other training institutions.

[a] Interviews with Huawei Technologies Co. Ltd, Beijing, 27 August 2021.
Source: Asian Development Bank.

Discussion

Deepening skills partnerships between VET institutions and enterprises has become a widely accepted policy alternative among the regions of the PRC. This type of TVET–industry linkage encourages local institutions to co-evolve with industrial partners and improves their capacity as sources of synthetic knowledge. Its experiences point to the possibility of rationalizing TVET programs with changing technology through close school–firm collaboration. Several key insights can be drawn from the PRC's case analysis.

First, the national policy framework can serve as political mobilization and an incentive plan for local adoption of skills partnership. On one hand, high-level authorities' commitment to rationalizing TVET through school–firm collaboration can mobilize regional support, because measurable school–firm linkages become a key performance indicator for local bureaucrats in political competition. On the other hand, the central government also attached monetary and nonmonetary incentives to its national strategy, including tax exemption, land use priority, and priority access to subsidized loans. Such political posturing and incentives have made the state's commitment to skills development credible for regional governments, VET institutions, and industry partners.

Second, identifying the appropriate skills partnership model is key for successful local implementation. A supportive national policy framework is a necessary but not sufficient condition for school–firm skills partnerships, and not all skills partnerships are equally effective in promoting school–industry linkage at the local level. During the PRC's transition from a low- to middle-income country, the parental and consortium models stood out as the most appropriate skills partnerships in manufacturing agglomeration regions. Both are deeply embedded in local economic, political, historical, and social networks and can empower local vocational colleges to provide state-of-the-art training for their industry partners, including large SOEs, SMEs, and high-tech multinational companies. In this regard, the national policy framework can create a conducive environment for policy adoption and diffusion, and the entrepreneurial approach can be embraced by local authorities, enterprises, and vocational institutions, which is the main driver for effective TVET–industry linkage.

Finally, the PRC experience verifies the argument that vocational education institutions can provide synthetic knowledge through their active and dynamic partnership with enterprises. The implementation of new manufacturing technologies demands a fundamental change in knowledge and competence of the current and future labor force, and thus ties industries with vocational education institutions (Lund and Karlsen 2020). The PRC's case demonstrates that government policies and measures for promoting TVET–industry linkage must focus on the creation of synthetic knowledge through dynamic partnership between TVET institutions and enterprises.

3.6. Contributions of the Qualifications System Despite the Absence of a Comprehensive Qualifications Framework for All Education and Training Programs

Background

Once skills are developed, qualifications play an important role in ensuring their social currency. The vocational qualification has multiple functions in the labor market. It can provide accurate information on an individual's competence for certain occupations. It also indicates each occupation's demand for particular types of skills and competencies. Moreover, the occupation-specific standards related to each qualification are important references for curriculum development in vocational education and training organizations and occupation activities carried out in the workplace. Given its close connection with a particular labor market, vocational qualification is highly contextualized and dynamic.

In advanced economies, the emergence of national qualifications frameworks is closely related to the competency-based approach to vocational education and the need to manage the proliferation of qualifications and qualification bodies that had developed from the 1960s (Corpus et al. 2007). As indicated in Chapter 1, the national qualifications framework is considered a building block in TVET and these are included in ADB projects executed in Cambodia, Indonesia, Mongolia, and elsewhere.

Prior analysis of types of national qualifications frameworks in Asia and the Pacific divided regional economies into five categories:

(i) a complete qualifications framework that covers all qualifications (vocational, training, and academic) such as Australia, New Zealand, and Malaysia;
(ii) a complete but not yet unified qualifications framework, including Hong Kong, China; and the Philippines;
(iii) a partial qualifications framework in which it is specific to type, level, or sector, such as Fiji, India, and Singapore;
(iv) a qualifications framework in development such as Brunei Darussalam, Pakistan, and Samoa; and
(v) not yet developed framework, including Bangladesh, Cambodia, and the PRC (Comyn 2009, Corpus et al. 2007).

In this regard, the PRC has no complete qualifications framework that covers all qualifications, and has effectively developed a binary qualifications framework consisting of an education qualifications system and an occupation qualifications system. The dual system was a legacy of the PRC's transition to the Socialist Market Economy (Fan and Liu 2012). Nationally, the Ministry of Education is in charge of educational qualifications and the Ministry of Human Resources and Social Security for managing the occupational qualifications system. Figure 7 illustrates the structure of the two qualifications systems.

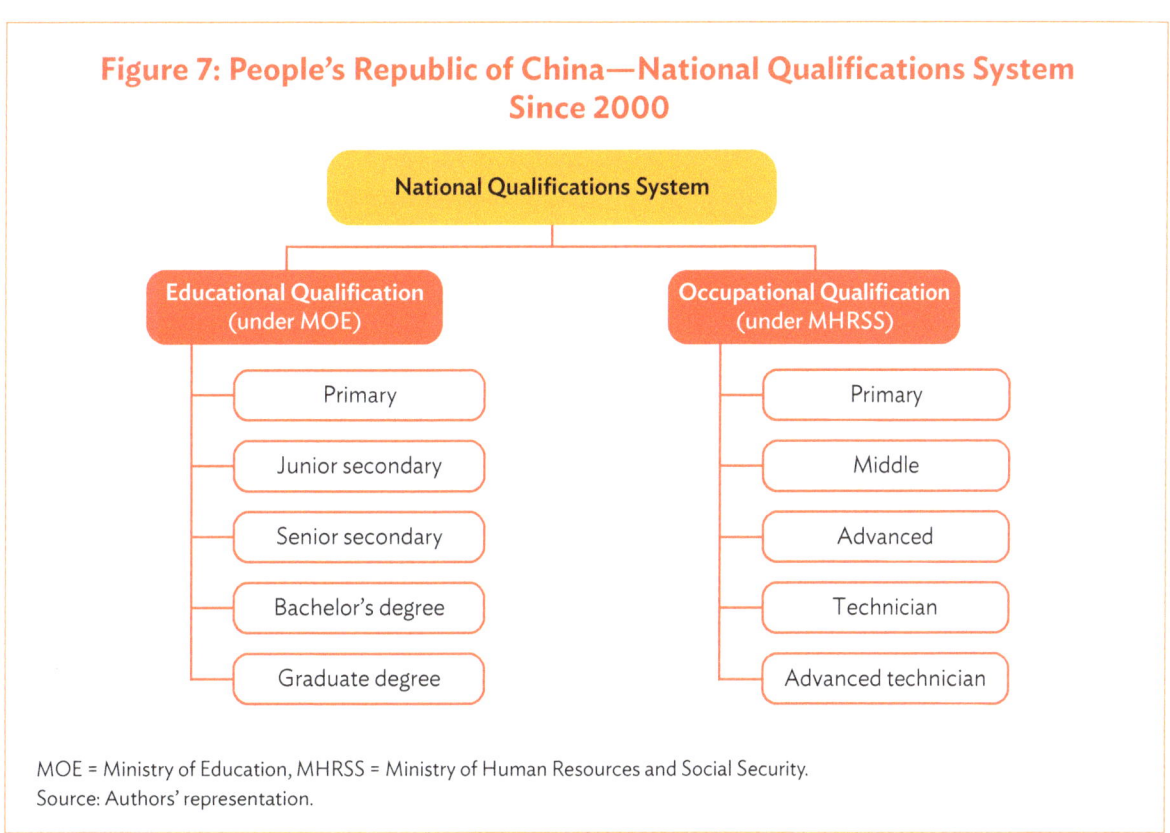

Figure 7: People's Republic of China—National Qualifications System Since 2000

MOE = Ministry of Education, MHRSS = Ministry of Human Resources and Social Security.
Source: Authors' representation.

Progress

(i) Accreditation of Qualification

The PRC is a newcomer in developing occupational classification systems. No explicit labor market occupation standard existed before 1949. In 1956, the State Council connected workers' compensation to their skill levels by adopting the Soviet-style worker assessment in the workplace. This workers' evaluation system stratified SOE workers into eight technical grades, which unified one occupation's technical grade and salary level. Grade certificates were awarded for "work type," which was defined by narrow occupational standards based on specific work tasks and types of equipment. This technical standard for workers was revised three times, in 1963, 1978, and 1989, reflecting changes in economic environment and industry demands (Chen 2000).

In 1999, the Ministry of Human Resources and Social Security (MHRSS) released the National Occupational Standards with reference to the Technical Procedures on Formulating National Occupational Standards (MHRSS 1999). A panel of 5–10 experts creates each set of occupational standards, composed of those who are familiar with the standard establishment methodology and with theoretical and industry experience. Standards for "white collar" and technical skills in intensive occupations (such as engineers and accountants) were developed and managed by the Division of Professional and Technical Personnel Management within the Ministry of Human Resources and Social Security. The new standards separated nonmandatory certificates, which should point to an individual's level of knowledge and technical skill, and mandatory certificates, which were preconditions for entry into common and technically complex work types related to public property, security, consumer interest, or health (Fang, Liu, and Fu 2009). National occupational standards are mandated to be updated every 3 to 5 years. Representatives from both industrial sectors and vocational education and training providers are invited to set up a committee in charge of updating national occupational standards. The establishment of National Occupational Standards sets the stage for the development of a national vocational qualifications system.

(ii) Occupational Skills Appraisal

The origin of the occupation testing system was the Soviet-style worker assessment in the 1950s. At that time, SOEs implemented independent evaluation of workers' occupation skills in the workplace, according to the Technical Level Standard for Workers of 1956, a decentralized approach to skills appraisal. In the early 1990s, the Ministry of Labor took a more centralized approach in occupation skills appraisal, issuing an administrative blueprint in 1993 for the extension of skills assessments beyond SOE workers, to the unemployed, students in schools and colleges, and other employees. The PRC's Labor Law (1994) provided the legitimacy for occupational standards, occupation skills testing, and an occupational qualification certificate system. The Technical Procedures on Formulating National Occupational Standards (MHRSS 1993) proposed to implement a socialized occupation testing system. The State Council made the Ministry of Human Resources and Social Security the supervisor of the newly created National Occupation Testing System in 1994.

The labor ministry further exerted its regulatory authority over skills certification through its National Occupational Testing Authority, which was accompanied by corresponding organs at the provincial level and in sector ministries (Figure 8). The Occupational Skills and Testing Authority was to appraise, coordinate, and monitor independent vocational skills appraisal stations, which could conduct the evaluation. They were also responsible for coordinating creation of new occupational standards and creating a national database of corresponding testing tasks.[32]

The operational capacity of the Occupational Skills and Testing Authority was substantially enhanced after the implementation of the World Bank's labor market project scheduled from 1996 to 1999. The project strengthened the authority's administrative capacity and facilitated the creation of occupational standards and the central database. The project helped set a methodological standard for developing occupational standards and test questions in the PRC. Following this methodological standard, various sector ministries issued their own standards in cooperation with the Occupational Skills and Testing Authority and produced more than 104 standards and standardized test questions for 49 standards by 2005 (Müller 2021).

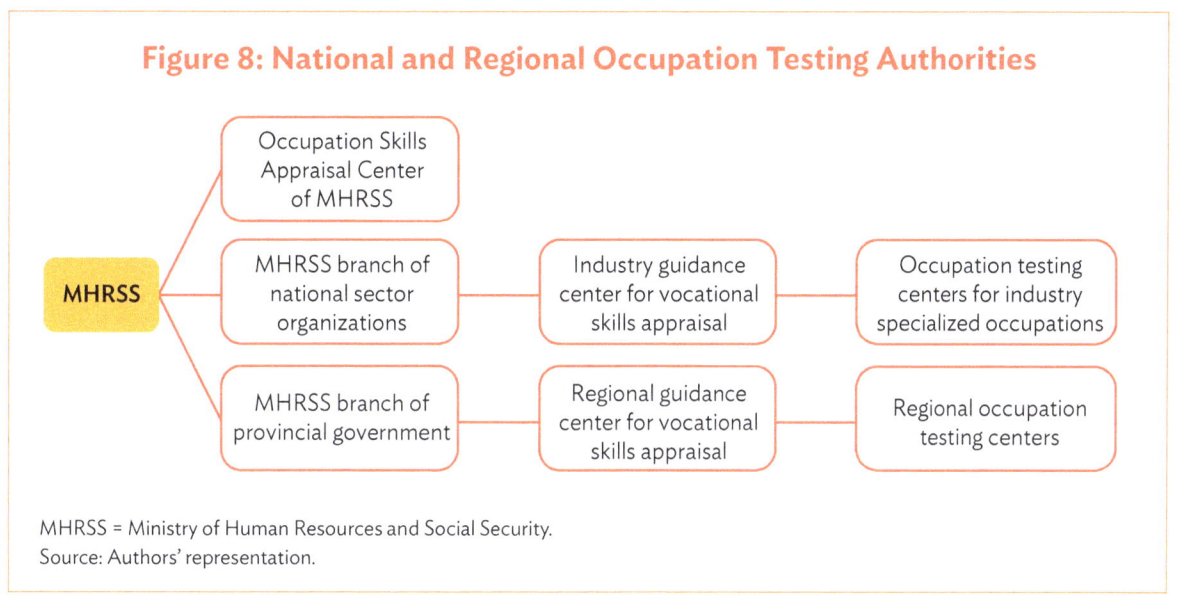

Figure 8: National and Regional Occupation Testing Authorities

MHRSS = Ministry of Human Resources and Social Security.
Source: Authors' representation.

(iii) Vocational Qualifications System

During labor market liberalization in the 1980s and 1990s, the government considered two options for vocational qualification: the German Dual System of Apprenticeship training, endorsed by the Education Commission and *Gesellschaft für Technische Zusammenarbeit*, the German government's implementing agency for technical development and cooperation; and the British National Vocational Qualifications, sponsored by Ministry of Labor, the British Embassy, and the British Council. In the late 1980s, the education ministry promoted learning from Germany's dual system. The return to economic reform in 1992 first shook up the cooperative arrangement with the

---

[32] Interview with a professor of Beijing Normal University, 16 August 2020.

Education Commission, and subsequently toward the Ministry of Labor developing its own system leaning toward British national vocational qualifications (Müller 2021). The new system expanded very quickly. For instance, the number of individuals participating in and passing the occupation testing rose in the previous 2 decades. Figure 9 shows the growth of vocational qualification certificates granted by the Occupational Skills and Testing Authority from 1996 to 2014.

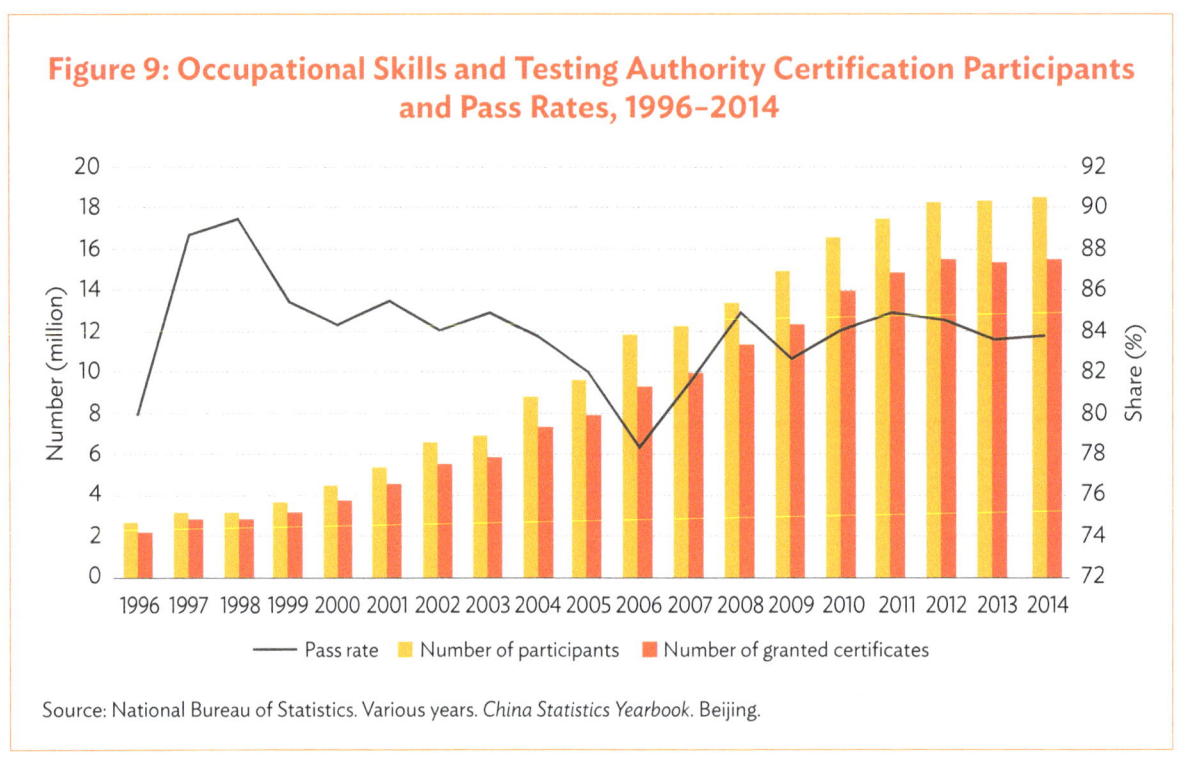

Figure 9: Occupational Skills and Testing Authority Certification Participants and Pass Rates, 1996–2014

Source: National Bureau of Statistics. Various years. *China Statistics Yearbook*. Beijing.

Eventually, the implemented vocational certificate system was a hybrid model, integrating elements of British national vocational qualifications' outcome-oriented certificates—with five hierarchical levels—with domestic policy elements in the form of occupational standards inherited from the Soviet model. The administrative structure and methods of skills appraisal under the PRC's national vocational qualifications also combined foreign and domestic practices. The interlacing system of vertical (sector) and horizontal (local) authority was extracted from the old Worker's Evaluation in SOEs. Meanwhile, the Occupational Skills and Testing Authority's functions of coordinating and monitoring skills appraisal stations within companies, vocational schools and colleges, and other training organizations resembled the British approach. In addition, the combination of theoretical test and practical skills assessment was the legacy of the Worker's Evaluation, but a company's skills assessment in the actual workplace was typical of the British national vocational qualifications.

Policy borrowing was also clear in the making of occupational standards for vocational qualification. Following the British model, the PRC's labor ministry perceived occupation standards as instruments to structure the curriculums of vocational training and bring them closer to market demand. This principle also manifested itself in training in school-based settings. The PRC vocational schools retained their discipline-based curriculums, and students were mandated to acquire a vocational certification before graduation.

Discussion

International comparative study claims that the PRC has yet to adopt a complete qualifications framework that covers all qualifications (Corpus et al. 2007). Alternatively, the country has adopted a hybrid vocational qualifications system with foreign and domestic characteristics. This National Vocational Qualification and its associated National Occupational Testing Authority and National Occupation Standards have provided accurate information on individual competence for certain occupations, and indicated each occupation's demand for particular types of skills and competencies.

Although a national qualifications framework has many obvious advantages—such as establishing commonality across different types of qualifications and specifying qualifications in terms of standards, levels, and outcomes—its establishment requires technical and institutional conditions. From the perspective of technical conditions for the national qualifications framework, the PRC is not ready for implementing a complete national qualifications framework because of the following limitations:

- *Due to the risk of poaching and labor market segregation, social demands for qualification were compromised.* Because the PRC's labor market is very flexible, most firms wish to hire skilled workers instead of training them. The PRC labor market is also segregated by its household registration system, so the vocational qualification cannot promote rural–urban mobility. Without the hope of becoming an urban resident by acquiring vocational qualification, migrant workers are reluctant to invest in their own training. About 60%–65% of migrant workers had only junior secondary education and 13%–17% of them had senior secondary education (Khor et al. 2016).

- *The current production regime does not support vocational qualification.* Given the PRC's long history with only a single academic qualifications system, vocational qualification is not considered as valuable as an academic degree. Enterprises' employment decisions depend mostly on academic degrees, not vocational qualification. In addition, the PRC's existing production regime does not support the growth of vocational qualification. The economic miracle of the past several decades was built on low-cost migrant workers and massive, standardized production. The competitiveness of firms depended critically on their low labor costs, not on the quality of their workforce. To address this problem, "Made in China 2025" is highlighting the transitioning to the production of high-end products, which means higher demand for skilled labor, especially highly skilled labor.

The development of national vocational qualifications in the PRC is further compromised by the lack of favorable institutional conditions. The country has neither adopted a competence approach to vocational education, nor "neoliberal" economic policies that emphasize the primary role of the private sector in economic development. Both were major drivers for a national qualifications framework in New Zealand and the UK in the 1990s but are currently unavailable in the country.

- *The PRC's academic qualifications system is hard to harmonize with vocational qualifications.* Academic degrees are based on analytical knowledge and vocational qualifications are built on synthetic knowledge. The PRC government enthusiastically encourages vocational schools to integrate industry needs and occupation standards with their discipline-based curriculums; however, such efforts are less likely to be successful due to the differences in knowledge base under the two types of qualifications.

- *The construction of a national qualifications framework implies centralization in the VET administration, but the PRC's qualifications systems are quite decentralized.* Various qualification bodies exist for different kinds of academic or vocational qualifications, which have competing interests and separate agendas. The introduction of a national qualifications framework will be challenged by qualification bodies.

- *Bureaucratic conflicts prevent further integration of national vocational qualifications.* After the organizational reforms of the State Council in 1988, 1998, and 2003, the majority of specialized economic ministries were consolidated to the National Development and Reform Committee. These reforms opened a window of opportunity to create a unified national qualifications system to replace the traditional system of vocational certification by line ministries. However, the central government has not adopted an encompassing national qualifications framework, due to bureaucratic conflicts between the education ministry and labor ministry (Müller 2021). In recent years, the intention has been to harmonize the two separate qualifications systems. In June 2014, the Modern Vocational Education System Construction Plan (2014–2020) stipulated that the PRC should develop a vocational education structure that ensures a smooth transition between secondary and higher levels within the vocational education system, and between vocational and the academic track. The Implementation Plan for the National Vocational Education Reform and Development (2019) further encouraged replacing the dual qualifications system with a new "1+X" system, under which vocational students will obtain one academic degree with multiple vocational competence certificates. The new policy will likely achieve its goal with more ministerial collaboration. For instance, the country established the National Inter-Ministerial TVET Coordination Committee under the State Council in 2018, which serves as the overall coordinating institution for TVET. The committee is led by the vice premier, with representatives from relevant ministries. The establishment of such a committee is an attempt to solve the problem of the lack of coordination.

The PRC's past experiences illustrate that a national qualifications framework was not a necessary or sufficient condition for a nation's transition to a middle-income country. When a country is competing with others on prices, not quality, qualifications and a national qualifications framework are not priorities for enterprises and industries. However, as the PRC now moves from middle-income to high-income, the creation of a national qualifications framework becomes a national imperative. The adoption of a complete national qualifications framework in the PRC depends on substantial changes in technical and institutional conditions.

In conclusion, from the PRC perspective, international experience can be a valuable reference for developing national qualifications frameworks, but the degree of policy adoption is influenced by historical and cultural factors, including but not limited to existing occupation standards, a skills appraisal system, a vocational qualification accreditation body, as well as alignment of different ministries' interests, current production regimes, labor market structure, orientation of vocation training, and an association between academic and vocational qualification. Without the support from these historical and cultural factors, the national qualifications framework is less likely to mobilize resources on the part of the government, the private sector, and the public.

3.7. Relationship Between the Development of General Education and Technical and Vocational Education and Training

Background

The PRC education system is divided into preschool, primary, secondary, and higher education, and into general and vocational tracks. In the general track, students start their primary education at 6 years old. After 6 years studying, they study for 3 years in junior secondary schools, and 3 years in senior secondary schools. Higher education for academic qualifications includes a bachelor's degree program and a postgraduate program. The vocational track consists of vocational education at primary, secondary, and higher levels. Secondary vocational education is provided by senior secondary vocational schools or technical schools. The higher levels of vocational education are undertaken by vocational colleges or by general institutions of higher learning (Fan and Liu 2012).

The PRC's general and vocational high schools are institutionally separated. Based on the Soviet model, the state has a dual education system at the secondary level, encouraging specialized education in separated systems. General and vocational education institutions differ in their educational missions, functions, curriculums, and in teaching, evaluation, quality assurance, and administration. General high schools provide preparatory education for college and are managed by the local education authority; vocational high schools offer preparatory training for employment and are managed by the education and labor authority. The social desirability for general and vocational high schools differs substantially. Parents in the PRC believe that academic high schools offer an opportunity for college education and white-collar jobs, while vocational schools accommodate "losers in academic competition" and lead to blue-collar jobs.

Progress

(i) Balanced Development of Upper Secondary Education

Since the early 1950s, the PRC had been purposely expanding its secondary vocational education to meet the skill demands of early industrialization. However, the Cultural Revolution interrupted formal schooling and workplace-based training from 1966 to 1976. To revitalize the economy, the PRC government carried out secondary education reform in the 1980s. Under the guidance of the State Council, the Ministry of Education proposed a balanced development trajectory for academic and vocational high schools through administrative regulation on schooling. These policies immediately took effect and changed the composition of secondary education schools (Figure 10).

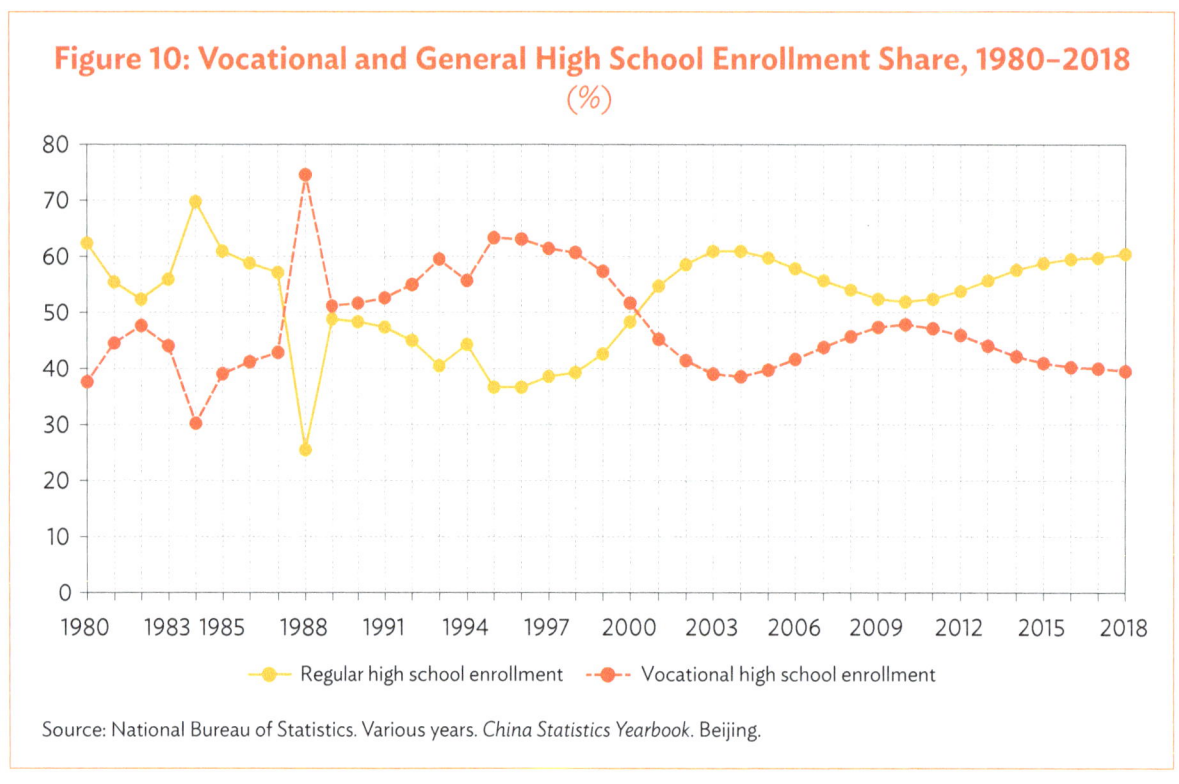

Figure 10: Vocational and General High School Enrollment Share, 1980–2018
(%)

Source: National Bureau of Statistics. Various years. *China Statistics Yearbook*. Beijing.

Keeping development of vocational and general high schools balanced is highly contentious in the PRC's contemporary society. Since 2000, the share of general high schools started to exceed vocational high schools. The difficulty in keeping an equal share of vocational schools was an unintended consequence of tertiary education expansion in the PRC. Starting 1999, the tertiary education expansion boosted social demand for general high school education and marginalized vocational high schools. The gross enrollment rate of tertiary education increased from 12.5% in 2000, to 26.5% in 2010, and 51.6% in 2019. As a result, the enrollment share of general high schools increased from 50% in 2000 to 60% in 2018.

(ii) Horizontal and Vertical Integration as a Policy Solution

When the PRC moves from an elite to a mass higher education system, the wage gap between college and high school graduates continues to grow (Khor et al. 2016). The tertiary expansion and the growing earning gap have forced the central government to reconsider its strategy for upper secondary education. As more students have aspired to receive higher education, vocational and general high school education have to be more flexible to accommodate student needs.

After much deliberation, the PRC government decided to establish a modern vocational education system in 2014, characterized by vertical and horizontal integration. Vertical integration refers to combining secondary vocational education with higher vocational education; horizontal integration refers to mobility between vocational and general high school education. The reform aimed at creating a "vocational education overpass" within the vocational education sector and between the vocational and academic education sector.[33]

Within the vocational education sector, the vocational education overpass involved creating integrated degree programs through collaboration of vocational high schools, vocational colleges, and universities of applied sciences. The detailed plan for vertical integration was introduced by a Decision of the State Council on Vigorously Promoting the Reform and Development of Vocational Education (2002), intending to establish a connection between secondary and tertiary vocational education systems. The vertical integration requires curriculum and teaching alignment at the secondary and tertiary levels, including harmonization of major categories, course content, occupation standards, and skills assessment and evaluation.

Several provinces launched pilot programs for vertical integration. For instance, Beijing started its "3+2" pilot in 2012, which included 3 years of education in vocational high school plus 2 years in vocational college. Thirty-four vocational high schools and 18 vocational colleges participated in this program. Shanghai's pilot program started in 2010 with only 4 vocational high schools and 3 vocational colleges. The program extended to 39 vocational institutions in 2013 and enrolled around 3,000 students. By 2014, Zhejiang Province enrolled 30,084 students in the vertical integration program, accounting for 13.5% of total vocational secondary enrollment.

On the horizontal dimension, the vocational education overpass refers to attempts to promote better integration of general and vocational high schools. Prior attempts can be categorized into three groups (Liu 2015).

The first one refers to on-campus curriculum exchange, where two types of high schools keep their own curriculum and borrow courses from each other. General high schools can adopt courses such as ICT from their vocational counterparts, and the two types of high schools remain independent organizations.

The second type is interschool curriculum cooperation. Under this model, general high school students can take courses in vocational ones and transfer their credits back, and vice versa. There are several pilots in Jiangsu, Zhejiang, and Hainan provinces, where general and vocational high schools collaborate in course development.

The last refers to comprehensive high school, which provides both academic preparatory course for college admission and vocational courses for employment. Beijing, Jiangsu, Heilongjiang, Hebei, Chongqing, Jilin, Sichuan, and Hunan provinces established comprehensive high schools, with 450 comprehensive high schools in the PRC by 2015.

---

[33] Interview with the director of Vocational Institution Development; director of Teaching and Quality Assurance, Bureau of Vocational and Adult Education, Ministry of Education, 28 August 2020.

Integrating general and vocational curriculums in comprehensive high schools has attracted much criticism. From the perspective of the institutional environment, this type of horizontal integration is hardly compatible with a student registration and management system as well as college entrance examination system. In addition, the funding formula, curriculum policy, and teacher preparation and professional development policy do not favor comprehensive high schools (Chang 2015).

Discussion

The vertical and horizontal integration of general and vocational education in the PRC faces serious challenges. First, such integration lacks legitimacy. The PRC government has initiated many pilots regionally without issuing formal legislation on integrating general and vocational education. It is hard to integrate education institutions under different ministerial controls. Vocational education at the secondary level is under the jurisdiction of the Ministry of Education and Ministry of Human Resources and Social Security, while most vocational colleges are under the control of the Ministry of Education. Hence, the vertical and horizontal integrations need cross-ministerial collaboration, which is easier said than done.

Second, integrating vocational and general education at the secondary level needs the help of a national qualifications framework, which can establish the equivalence of vocational degrees or certificates and academic degrees. The PRC has yet to adopt a complete national qualifications framework, as noted, without which it is hard to change social perception of vocational education and promote student mobility across sectors.

Third, government incentive for innovation is insufficient. In the PRC context, local education authorities are often measured by their rankings on college entrance examinations, not by the production of skilled workers. Although vertical and horizontal integration was endorsed by the State Council of the People's Republic of China, it can hardly change the incentive structures facing local bureaucrats.

Fourth, tension exists between balancing development of vocational and general education and the goal of horizontal integration. Balanced development implies keeping the current dual track system, which makes it literally impossible to bridge the two sectors through horizontal integration.

Finally, integration of vocational and general education at the tertiary level is likely to replace integration at the secondary level. The latest tertiary expansion has fundamentally changed the landscape of the PRC education sector. Even academically less prepared high school students can easily be recruited by vocational colleges through open admission. This implies that sorting into a vocational track will be postponed until college. Thus, the ideas of keeping a dual track high school system will become obsolete and irrelevant. Given the trend of tertiary education expansion globally, the PRC's attempts at integrating vocational and general education tracks at the tertiary level may become a norm rather than an exception in the near future.

Chapter 4
CONCLUSION

4.1 Lessons Learned

These chapters have examined approaches used broadly within international development of TVET, specifically focusing on suitability within the context of the developing world. The building-block approach that has emerged in the latest era of TVET development is clearly based on sound principles; however, this study finds much evidence to suggest that it may not necessarily be the best fit for developing nations. A "one-size-fits-all" approach seems doomed by the unique circumstances of each nation, which play a significant role in the efficacy of any policy initiative. The most effective solutions should instead be constructed around the existing environment of developing nations, as is highlighted through the case studies of the PRC and the Republic of Korea. Through the study of these cases of development, three key lessons for TVET policy emerge pertaining to the role of the state and private sector involvement.

(i) Role of the State

The Republic of Korea and the PRC provide examples of TVET development models based largely on state involvement, as opposed to liberal, market-driven policy approaches. In the Republic of Korea, an authoritarian government was able to enforce training mandates on large corporations to increase industry involvement in the skills formation process. This policy was heavily influenced by the circumstances of the time. First, the withdrawal of US troops from the Korean peninsula would mean that the Republic of Korea would look to heavy industrial manufacturing as a way of boosting national defense. Second, during this period, the powers of President Park Chung-hee have been expanded; thus, dissent from private enterprises could be somewhat ignored. The success of a state-led approach is not only related to the extent to which the government can act flexibly according to its own mandate, but also depends on the quality of the civil service handling these affairs. The long-standing esteem for civil service careers in the Republic of Korea filled the civil service with some of the most capable people in the country. It is not difficult to understand how high-quality civil service employees can be crucial where there is a high degree of state autonomy.

The focus of the Republic of Korea's leadership was also very much directed toward human capital development, due to the lack of natural resources that could be used to boost economic growth. As a result, skills forecasting was carried out more through target-setting than any kind of projected analysis, and there was a relatively high degree of investment in vocational schools, including some elite vocational schools that were able to draw a talented pool of students who could not afford general schooling. The strong state focus on education and human capital early on in national development has been cited as one of the key factors in country's success. This was a state-led initiative that required a strong authoritarian-style mobilization of resources, and thus the ability to pursue integrated development strategies such as this can be viewed as the strength of a state-led approach.

The benefits of a state-led approach can also be seen throughout the history of TVET development in the PRC. Early industrialization hinged on the ability of SOEs to be involved in training and to be obliged to provide lifetime employment for their employees. Through the proactive approach to modeling and piloting led by the PRC government, a potential solution is offered to the innovation bottleneck typically associated with state-led approaches. In the PRC, the importance of buy-in at the highest levels of government cannot be understated, as TVET policy (and all other forms of policy) were framed within the bigger picture of national development.

Clearly, there are advantages to the state-led approach that may be useful to other developing nations. Top-down leadership afforded the PRC and the Republic of Korea the ability to rapidly mobilize resources to support ambitious, widespread national development initiatives, and institutions and policy could then be iterated according to specific needs. In the long term, it is likely necessary for the state to relinquish at least some control over the system to allow for demand-driven growth, but in the earlier stages where institutions and the private sector are immature, state-led directives can help establish a firm, positive direction to get development on track.

(ii) Private Sector Involvement

The involvement of the private sector in the TVET system is clearly one of the most important factors in its success. Early on, the Republic of Korea's industrial sector was so small and undeveloped that there were not enough experts who could be consulted on skills standards creation. Also, the focus was on low-skill manufacturing jobs, so it was difficult to find people with experience who could also advise the government. As a result, standards creation was state-led through a commissioned group of experts, most of whom were professors and high school teachers. This was not a perfect approach, but inevitably would be changed with the development of a more demand-led and systematic approach to standards creation—although this shows how a nascent, underdeveloped private sector may not be ready to take on key roles within the likes of sector skills councils immediately. The PRC experience also echoes this notion, wherein the internal planning mechanisms of the government were used to direct skills development in the earliest stages, and even throughout the liberalization of the labor market, external planning mechanisms that focused on specialized economic ministries, regional governments, and "sector organizations" were the key agents within the PRC skills development system. These approaches highlight the unsuitability of sector skills councils in skills development when the private sector is immature and unspecialized, and potential alternatives can be utilized.

Within the PRC case, there are also lessons on how policy should target the private sector for TVET. It is important that policy considers the type of skills formation regime within the target environment, as needs can change drastically based on such differences.

The early approach to TVET in the 1950s and 1960s in the PRC can be considered as something resembling a collective skill formation regime. There was no shortage of industry involvement in training; instead, the risk might be that there was too much. As training was effectively run through the SOEs, there was no disconnect between training and the needs of industry, but the skills imparted in training may have been so specific as to render skills acquired useless outside of their own very specific role.

Later on, in the social market economy era, the move toward statist skills formation regime meant a distinct absence of industry within the TVET system and some mechanism for sourcing industry influence on training and curriculums needed to be sought. The important point is that the demand and supply of the PRC TVET system were vastly different at these two stages of development, thus it is difficult to reconcile how a "one-size-fits-all" approach could possibly capture the needs of developing nations all along the development spectrum. The modern PRC TVET approach demonstrates the potential for the use of PPPs, but also highlights the risks of this approach in funding shortages for TVET.

(iii) Technical and Vocational Education and Training Within the Broader Educational System

Another clear lesson from examination of these cases is the need to consider TVET development in relation to general education. The Republic of Korea paid much attention to educational expansion following the Korean War, and made remarkable strides in expanding basic primary education. This has a huge complementary effect on TVET education, as students are equipped with basic literacy and numeracy skills. However, the knock-on effect of this was the subsequent expansion of secondary education, which pulled in more students than vocational schools due to the generally higher esteem for general education. However, fee exemptions meant that TVET could attract a group of intelligent students who could not afford general schooling, which would boost the overall outcomes of TVET education. However, once university education was expanded in the 1990s, the numbers for TVET began to dwindle rapidly, going from roughly 50:50 to almost 70:30. The knock-on effects this would have for TVET funding would be very significant. This case is not all that dissimilar to the PRC, where the expansion of university education preceded a precipitous fall in the percentage of students attending TVET schools.

It is clear that in countries like the PRC and the Republic of Korea, where a pre-established high esteem for general education exists, it is very important to consider the concurrent development of general education alongside TVET. Both countries have a long history of veneration of academic education, where the Confucian ideals of intellectualism are tied to social prestige and even patriotism. In countries where there is a predetermined preference for one type of education over another, perhaps policy makers may need to figure out some kind of mechanism for limiting the number of students who may attend each one. The PRC has looked to vertical and horizontal integration as a way of attracting people into the TVET track, whereby students in the vocational track could eventually pursue some kind of university education or be somehow affiliated with general education schools. In the PRC, vertical and horizontal integration efforts have been challenging, for a number of reasons.

One critical issue concerning the regard for TVET education is the degree to which qualifications have value on the labor market. When employers are aware of what certain qualifications mean and trust their validity, then more students are encouraged to pursue that qualification. The modern solution to this problem is the aforementioned national qualifications framework, which prescribes and describes the contents and outcomes of training, for students, trainers, and employers alike.

However, the PRC and the Republic of Korea both provide examples that may undermine the value of the national qualifications frameworks in the earlier stages of development. In the earliest stages of development, the Republic of Korea began with a completely unregulated qualifications network, and would eventually roll out a very minimalist framework that contained standards for a limited amount of professions. While this system was not perfect, a more comprehensive framework was not urgently needed primarily because most of the jobs were very low-skilled and qualifications were not generally required. As the Korean economy developed and jobs became more specialized, there was a need to roll out a national testing system that would apply to all the various defined professions, but this was not a comprehensive national framework by any means, with various professions still managed entirely by specific sector ministries.

There are also huge mobilization costs associated with creating a national qualifications framework, due to the systematic incongruencies of developing nations. For example, the PRC's efforts to develop a national qualifications framework were deterred due to an immature labor market filled with unskilled workers, an unwillingness of enterprises to train employees due to the risk of poaching, the lack of any suitable central authority to regulate the system, and other factors. Both the PRC and the Republic of Korea would eventually move toward something resembling a national qualifications framework, but their examples highlight the potential for rapid skills development without a comprehensive framework, particularly in the earlier stages. Thus, from the perspective of developing countries facing the most basic challenges of TVET development, a national qualifications framework may be a costly measure that does not entirely suit their short-term needs.

4.2 Final Note

In the early stages of international development, developed nations played a more paternalistic role in international development and thus solutions were generally developed through western expertise and with less consideration for the agency of developing nations. As time has gone on and the flaws of ignoring the perspective and expertise of aid-recipient nations have become clear, the relationship between donor and recipient has become increasingly based on mutual participation. This change can demonstrably be seen through the likes of the 2005 Paris declaration, but questions still remain over the degree to which it has been implemented. Certainly, such a drastic shift cannot happen overnight, and even if there is no intentional effort to preserve an international agenda wherein developed nations have the loudest voices, there are many different ways through which such a dynamic could be preserved unintentionally.

Specifically, in TVET, the two countries that have most dominated the attention of practitioners and academics have been Germany and the United Kingdom. The German Dual System has traditionally been seen as the most effective national TVET system, with German TVET graduates demonstrating particularly high achievement and TVET being relatively popular among the public. More recently, the British National Qualifications Framework has been viewed as a solution that may help systematically organize a previously somewhat unregulated TVET environment. The legacy of the admiration of these TVET approaches has been the regular inclusion of PPPs, sector skills councils, competency-based curriculums, and more within TVET programming in aid-providing organizations.

Despite their theoretical benefits, there are numerous reasons to question the suitability of these systems as a model for developing nations. For one, the inclusion of features from both models can potentially be contradictory, as Germany and the UK are often considered virtual opposites within the academic discussions around TVET system typologies. Germany is considered a collective skills formation regime, where a high degree of cooperation exists between the private sector and the state in the organization and running of the TVET system. Conversely, the UK is considered a liberal skills formation regime, where there is little state or private sector involvement in the organization and running of the TVET system. TVET policy within these two vastly different environments serves to address entirely distinct systematic needs, which leaves many questions as to how they could be used alongside one another. Furthermore, many other different types of TVET systems have their own specific areas of need. The statist skills formation regime of France features high state commitment but a lack thereof from industry, meaning that a well-resourced public TVET system may lack the necessary input of industry as far as competencies and skills planning are concerned. The segmentalist skills formation regime is marked by the opposite, with a high degree of employer involvement in training and lack of regulation from the state, as is seen in the likes of Japan. While these typologies are not the most nuanced way of describing TVET systems, they do offer a convenient lens to view and describe the variety of systems and how their policy needs depend on the different approaches governments have taken.

This contradiction speaks to a broader issue within the discussion around TVET in international development, which is the lack of consideration for the unique qualities of each TVET system and the unique factors which have led to their emergence. Looking at the examples within Europe alone, countries like France, Germany, and the UK can all trace their TVET origins back to a relatively homogenous guild system, yet the way in which these countries responded to significant historical events eventually resulted in an abundance of TVET diversity, even within the very same region and to the present day. Not enough consideration has been given to the relationship between the regional environment and the TVET system, which plays a crucial role in their formation and continuation. The transfer of best-practice solutions to new contexts has been wrought with many difficulties, and it is very possible that these difficulties are precisely a consequence of the lack of consideration for the unique regional circumstances of each developing nation.

The examples of the PRC and the Republic of Korea, as noted, present clear evidence that success in TVET development is possible without adherence to best-practice solutions in Europe, and that there is a strong case for incremental policy that addresses specific needs of development. Through the PRC and the Republic of Korea, we can see that TVET development has been subjected to a confluence of seemingly unrelated issues, including national security, cultural philosophy, leadership style, and other areas, highlighting the many different potential areas of divergence that could render the needs of one environment completely different from another. These examples serve to support the notion that no two cases can reasonably be considered the same within international development, and that it is crucial that every case be evaluated according to specific factors.

The implications of this knowledge are that approaches based on the experiences of foreign nations should not be transferred without a specific examination of the environment of the aid-recipient country. And that flexibility should be applied in all cases where specific features from foreign models are included with conditionalities in developmental programming.

This is not to suggest no building blocks can be taken from successful contexts and prescribed to developing contexts. In the German TVET system, it can be seen that a strong training culture among private partners can be extremely conducive to effective TVET. This, alongside reliable assessment systems, strong TVET leadership, and appropriate investment in necessary equipment are all prescriptions based on quality, as opposed to features attached to specific methods of conducting TVET. In these quality-based prescriptions, consideration of the system of origin and how it may differ from the particular developing context is still needed, but it is likely that building blocks such as these can provide strong examples in areas of importance that are, to a certain extent, context-agnostic.

The findings of this study need not be specifically contained to TVET, as this is just one example of policy that is largely influenced by existing institutions and unique historical events. In fact, it is quite likely that these findings would apply to all areas of policy making, especially those areas that have long been subject to the attention of social policy, such as education.

REFERENCES

Abel, H. and H. H. Groothoff. 1959. *Die Berufsschule. Gestalt und Reform.* Darmstadt: Carl Winter.

Abbot, A. 1933. *Education for Industry and Commerce in England.* London: Oxford University Press.

Asian Development Bank (ADB). 2008. *Public–Private Partnership Handbook.* Manila: ADB.

_____. China, People's Republic of: Shanxi Technical and Vocational Education and Training Development Demonstration Project. https://www.adb.org/projects/51382-002/main.

_____. Georgia: Modern Skills for Better Jobs Sector Development Program, Subprogram 1. https://www.adb.org/projects/52339-001/main.

_____. India: Himachal Pradesh Skills Development Project. https://www.adb.org/projects/49108-002/main.

_____. India: Madhya Pradesh Skills Development Project. https://www.adb.org/projects/48493-001/main.

_____. Nepal: Skills Development Project. https://www.adb.org/projects/38176-012/main.

_____. Timor-Leste: Mid-Level Skills Training Project. https://www.adb.org/projects/45139-001/main.

_____. 2011a. *Project Preparatory Technical Assistance Report for Skills Development Project in Nepal.* Manila.

_____. 2011b. *Report and Recommendation of the President to the Board of Directors: Proposed Grant to the Democratic Republic of Timor-Leste for the Mid-Level Skills Training Project.* Manila.

_____. 2014a. *Report and Recommendation of the President to the Board of Directors: Proposed Loan to the Kingdom of Cambodia for the Technical and Vocational Education and Training Sector Development Program.* Manila.

_____. 2014b. *Report and Recommendation of the President to the Board of Directors: Proposed Loan to Mongolia for the Skills for Employment Project.* Manila.

_____. 2016a. *Completion Report: People's Republic of China: Chongqing Vocational Training Information Management Platform Development.* Manila.

_____. 2016b. *Project Preparatory Technical Assistance Report for Madhya Pradesh Skills Development Project in India.* Manila.

_____. 2017. *Report and Recommendation of the President to the Board of Directors: Proposed Loan to India for the Himachal Pradesh Skills Development Project Distribution.* Manila.

_____. 2018a. *Concept Paper: Proposed Grant to Tajikistan for the Skills and Competitiveness Sector Development Program.* Manila.

_____. 2018b. *Report and Recommendation of the President to the Board of Directors: Proposed Loan to the Republic of Indonesia for the Advanced Knowledge and Skills for Sustainable Growth Project.* Manila.

_____. 2018c. *Technical Assistance to the People's Republic of China for Shanxi Technical and Vocational Education and Training Development Demonstration Project.* Manila.

Asheim B. T. 2012. The Changing Role of Learning Regions in the Globalizing Knowledge Economy: A Theoretical Re-Examination. *Regional Studies*. 46 (8). pp. 993–1004.

Ashton, D. et al. 1999. *Education and Training for Development in East Asia: Political Economy of Skill Formation in the Newly Industrialized Economies.* Milton Park, Oxfordshire: Routledge.

Ashton, D. et al. 2002. The Evolution of Education and Training Strategies in Singapore; Taipei,China; and South Korea: A Development Model of Skill Formation. *Journal of Education and Work*. 15 (1). pp. 5–30.

Balogh, T. 1964. The Economics of Educational Planning: Sense and Nonsense. *Comparative Education*. 1 (1). pp. 5–17.

Becker, G. 1964. *Human Capital.* Chicago: University of Chicago Press.

_____. 1993. *Human Capital. A Theoretical and Empirical Analysis, with Special Reference to Education.* 3rd ed. Chicago: University of Chicago Press.

Blackbourn, D. 1977. The Mittelstand in German Society and Politics. *Social History*. 4. pp. 409–433.

Blankertz, H. 1982. *Die Geschichte der Padagogik: Von der Arifkliirwzg his zur Gegenwart.* Wetzlar: Büchse de Pandora.

BiBB. 2014. *Datenreport zum Berufsbildungsbericht.* Bonn: Bundesinstitut für Berufsbildung.

Britannica. *Public–Private Partnership.* https://www.britannica.com/topic/public-private-partnership.

Brockmann, M., L. Clarke, and C. Winch. 2011. *Knowledge, Skills and Competence in the European Labour Market.* New York: Routledge.

Busemeyer, M. R. 2015. *Skills and Inequality: Partisan Politics and the Political Economy of Education Reforms in Western Welfare States.* Cambridge: Cambridge University Press.

Busemeyer, M. R. and C. Trampusch. 2012a. Introduction: The Comparative Political Economy of Collective Skill Formation. In M. R. Busemeyer, and C. Trampusch, eds. *The Political Economy of Collective Skill Formation.* London: Oxford University Press.

———. 2012b. *The Political Economy of Collective Skill Formation.* London: Oxford University Press.

Busemeyer, M. R., and T. Iversen. 2014. The Political Economy of Skills and Inequality. *Socio-Economic Review.* 12 (2). pp. 241–243.

Carbonnier, G., M. Carton, and K. King. 2014. International Education and Development: Histories, Parallels, Crossroads. *International Development Policy.* http://journals.openedition.org/poldev/1767.

Chang, B. N. 2015. Current Status, Issues and Solutions for Chinese Comprehensive High School Development. *Education Development Research.* 33 (2). pp. 69–74.

Chen, L. 2008. *Competence, Curriculum and Qualification: From Work and to Work.* Beijing: China Labor Press.

Chen, L. X. 2000. *Fifty Years of Chinese Vocational Training.* Beijing: China Employment Training Technical Instruction Center.

Chen, S. S. 2012. Evolution of Chinese Public Policy Mode from the Perspective of Interests Group. *Social Science and Medicine.* 8. pp. 13–23.

Chen, Y. 2009. *Skills Revitalization: Strategy and Technology.* China Labor and Social Security Press.

Choi, Y., H. Oh, and Y. Choi. 2009. *A Study on Establishing a Cooperation Model for Korean Vocational Education and Training Development.* Seoul: Ministry of Education, Science and Technology.

Choi, Y. and H. Choi. 2004. *Estimation of Appropriate Size of Vocational Education and Improvement of Vocational Education System.* Seoul: Korea Institute for Economics and Trade.

Comyn, P. 2009. Vocational Qualification Frameworks in Asia-Pacific: A Cresting Wave of Educational Reform? *Research in Post-Compulsory Education.* 14 (3). pp. 251–268.

Corpus, M. et al. 2007. *Qualification Frameworks in the Asia-Pacific Region.* https://www.apqn.org/media/project_group_reports/pg2_project_report_oct_2007.pdf.

Danforth, B. 2014. *The Development of Education Systems in Advanced Capitalist Societies.* PhD Dissertation. University of North Carolina at Chapel Hill.

Deissinger, T. 1994. The Evolution of the Modern Vocational Training Systems in England and Germany: A Comparative View. *A Journal of Comparative Education.* 24. pp. 17–36.

Deissinger, T. 2002a. Apprenticeship Systems in England and Germany: Decline and Survival. *Toward a History of Vocational Education and Training (VET) in Europe in a Comparative Perspective.* Florence: European Centre for the Development of Vocational Training.

_____. 2002b. Different Approaches to Lifelong Learning in Britain and Germany: A Comparative View with Regard to Qualifications and Certification Frameworks. In K. Harney, ed. *Lifelong Learning: One Focus, Different Systems.* Frankfurt: Lang.

_____. 2015. The German Dual Vocational Education and Training System as "Good Practice"? *Local Economy.* 30 (5). pp. 557–567.

Doner, R. F., and B. R. Schneider. 2016. The Middle Income Trap: More Politics than Economics. *World Politics.* 68 (4). pp. 608–644.

Easterly, W. 2002. *The Elusive Quest for Growth: Economists' Adventures and Misadventures in the Tropics.* Cambridge: MIT Press.

Economic Planning Board. 1961. *Survey Report on the Employed Technology Human Resources.* Seoul.

Fan, W., and Y.F. Liu. 2012. Overview of China's Qualification System. *Policy Notes for Promoting Skills Development and Job Creation in East Asia Project.* http://documents1.worldbank.org/curated/en/216651508745207540/pdf/120583-WP-P150980-PUBLIC-China-national qualification framework-summary.pdf.

Fang, Z. H., H. Liu, and X. L. Fu. 2009. *Knowledge and Skill: 60 Years of Chinese Vocational Education.* Zhejiang University Press.

Fawcett, C., G. El Sawi, and C. Allison. 2014. *TVET Models, Structures, and Policy Reform: Evidence from the Europe and Eurasia Region.* Washington, DC: United States Agency for International Development.

Fortwengel, J. 2017. Practice Transfer in Organizations the Role of Governance Mode for Internal and External Fit. *Organizational Science.* 28 (4). pp. 895–909.

Fortwengel, J., and G. Jackson. 2016. Legitimizing the Apprenticeship Practice in a Distant Environment: Institutional Entrepreneurship through Inter-Organizational Networks. *Journal of World Business.* 51 (6): pp. 895–909.

Foster, P. 1965. The Vocational School Fallacy in Development Planning. In A. A. Anderson and M.J. Bowman, eds. *Education an Economic Development.* Chicago: Aldine Publishing.

Glaeser, E. L. and M. Lu. 2018. Human-Capital Externalities in China. *NBER Working Papers.* Cambridge, MA.

Green, F. et al. 1999. The Role of the State in Skill Formation: Evidence from the Republic of Korea; Singapore; and Taipei,China. *Oxford Review of Economic Policy*. 15 (1). pp. 82–96

Gow, J. I., and C. Dufour. 2000. Is the New Public Management a Paradigm? Does It Matter? *International Review of Administrative Sciences*. 66 (4). pp. 573–597.

Hall, P. A., and D. Soskice. 2001. An Introduction to Varieties of Capitalism. In P. A. Hall and D. Soskice, eds. *Varieties of Capitalism: The Institutional Foundations of Comparative Advantages*. London: Oxford University Press.

Han, F. 2016. Promoting Government-Social Partner Collaboration in the Construction of Modern Vocational Education System. *Reviews of Economic Research*. 61. pp. 3–17.

Hollander, A. and N. Y. Mar. 2009. Toward Achieving TVET for All: The Role of the UNESCO-UNEVOC International Center for Technical and Vocational Education and Training. In *International Handbook of Education for the Changing World of Work*. pp. 41–57. Dordrecht: Springer.

Hoogenboom, M. et al. 2018. Guilds in the Transition to Modernity: The Cases of Germany, United Kingdom, and the Netherlands. *Theory and Society*. 47. pp. 255–291.

Huawei. Huawei ICT Academy: Building a Talent Ecosystem and Boosting the ICT Industry's Development. https://e.huawei.com/za/publications/global/ict_insights/201907041409/talent-ecosystem/huawei-ict-academy.

Human Resources Development Service of Korea. 2002. *20 Years History of Human Resources Development Service of Korea*. Seoul.

International Bureau of Education. Competency-Based Curriculum. http://www.ibe.unesco.org/en/glossary-curriculum-terminology/c/competency-based-curriculum.

International Labour Organization (ILO). 2016. *Using Labor Market Information*. Geneva.

_____. 2019. *Sector Skills Council: Policy Note*. Jakarta.

_____. 2021. Public–Private Partnerships. https://www.ilo.org/pardev/public-private-partnerships/lang--en/index.htm.

Jeong, T. 2008. *Changes and Tasks in the Vocational Competency Development System*. Seoul: Korea Vocational Competency Development Institute.

Ji, M. 2014. A Study on the Life History of Skilled Workers in the Early Stages of Heavy and Chemical Industrialization. Master's Thesis in Sociology. Yonsei University.

Khor, N. et al. 2016. China's Looming Human Capital Crisis: Upper Secondary Educational Attainment Rates and the Middle-income Trap. *China Quarterly*. 228. pp. 905–926.

Kieseweiter, H. 1989. *Industrielle Revolution in Deutschland, 1815–1914*. Frankfurt: Suhrkamp.

Kim, H. 1999. A Study on Youth Problems in the Japanese Empire. PhD Thesis in Sociology. Yonsei University.

Kim, S. and J. Sung. 2005. *Employment Policy in Korea*. Seoul: Korea Labor Institute.

King, K. 2009. A Technical and Vocational Education and Training Strategy for UNESCO. A Background Paper. Unpublished background paper for the International Expert Consultation Meeting on Technical and Vocational Education. January, 2009. Paris.

King, K. and C. Martin. 2002. The Vocational School Fallacy Revisited: Education, Aspiration and Work in Ghana 1959–2000. *International Journal of Educational Development*. 22. pp. 27–28.

Kliebard, H. M. 1999. *Schooled to Work: Vocationalism and the American Curriculum, 1876–1946*. New York: Teachers College Press.

Kümmel, K. 1980. Zur Schulischen Berufserziehung im Nationalsozialismus (Vocational Education in School under National Socialism). In M. Heinemann, ed. *Erziehrmg und Schulwrg im Drillen Reich, Teil 1: Kindergarten, Schule, Jugend. Berufserziehtmg*: pp. 275–288. Stuttgart: Klett-Cota.

Lauder, H., P. Brown, and D. Ashton. 2017. Theorizing Skill Formation in the Global Economy. In J. Buchanan et al., eds. *The Oxford Handbook on Skills and Training*. Oxford: Oxford University Press.

Lewis, T. 2007. The Problem of Cultural Fit—What Can We Learn from Borrowing the German Dual System? *Compare: A Journal of Comparative and International Education*. 37 (4). pp. 463–477.

Lewis, W. A. 1954. Economic Development with Unlimited Supplies of Labour. *Manchester School of Economic and Social Studies*. 22 (2). pp. 400–449.

Lieberthal, K. G., and M. Oksenberg. 1998. *Policy Making in China: Leaders, Structures, and Processes*. Princeton, New Jersey: Princeton University Press.

Lim, S. 2015. Establishment of Kumoh Technical High School and the Utilization of Elite Skilled Manpower. Master's Thesis in History of Science and Philosophy of Science. Seoul National University.

Liu, L. Q. 2015. Reflection and Policy Recommendation for Connecting General and Vocational Education at High School Level. *Education Research*. 9. pp. 92–98.

Lund, H. B., and A. Karlsen. 2020. The Importance of Vocational Education Institutions in Manufacturing Regions: Adding Content to a Broad Definition of Regional Innovation Systems. *Industry and Innovation*. 27 (6). pp. 660–679.

Lundvall, B. 2010. *Introduction to National System of Innovation National Systems of Innovation: Toward a Theory of Innovation and Interactive Learning.* London and New York: Anthem Press.

Marques, I. 2017. *Political Connections and Non-traditional Investment: Evidence from Public–Private Partnerships in Vocational Education.* Moscow: Department of Political Science. National Research University–Higher School of Economics.

McGrath, S. A. 2012. Vocational Education and Training for Development: A Policy in Need of a Theory? *International Journal of Educational Development.* 32 (5). pp. 623–631.

Ministry of Education and Research. 2015. *Report on Vocational Education and Training 2015.* Beijing: Federal Ministry of Education and Research.

Mittmann, F, 1998. Por Necesidad y Pro Moda. *La Nacion* (San Jose, Costa Rica). 26 March.

Müller, A. 2021. Bureaucratic Conflict between Transnational Actor Coalitions: The Diffusion of British National Vocational Qualifications to China. *Social Policy and Administration* 55 (1). pp. 1021-1035.

National Bureau of Statistics. Various years. *China Statistics Yearbook.* Beijing.

Nelson, R. R. 1994. The Co-evolution of Technology, Industrial Structure, and Supporting Institutions. *Industrial and Corporate Change.* 3 (1): pp. 47–63.

Ngcwangu, S. 2015. The Ideological Underpinnings of World Bank TVET Policy: Implications of the Influence of Human Capital Theory on South African TVET Policy. *Education as Change.* 19 (3). pp. 24–45

Oakes, J. 1985. *Keeping Track: How Schools Structure Inequality.* New Haven: Yale University Press.

Okwuanaso, S. 1984. The Fallacy of Vocational Education in Developing Nations. *Canadian Vocational Journal.* 20 (1). pp. 16–18.

Paik, S. 2013. 2012 Modularization of Korea's Development Experience: Role of Private Schools in Korea's Educational Development. Seoul: KDI School of Public Policy and Management.

Perry, P. J. C. 1976. The Evolution of British Manpower Policy from the Statute of Artificers 1563 to the Industrial Training Act 1964. London: The British Association for Commercial and Industrial Education (BACIE).

Psacharopoulos, G. 1991. Vocational Education Theory, VOCED 101: Including Hints for "Vocational Planners". *International Journal of Educational Development.* 11. pp. 193–199.

____. 2006. World Bank Policy on Education: A Personal Account. *International Journal of Educational Development.* 26 (3). pp. 329–338.

Psacharopoulos, G. and M. Woodhall. 1985. *Education for Development: An Analysis of Investment Choices*. New York: Oxford University Press.

Psacharopoulos, G. 2006. World Bank Policy on Education: A Personal Account. *International Journal of Educational Development*. 26(3). pp. 329-338.

Remington, T. F. 2016. Regional Variation in Business-Government Relations in Russia and China. *Problems of Post-Communism*. 2. pp. 1–12.

_____. 2017. Business-Government Cooperation in VET: A Russian Experiment with Dual Education. *Social Science Electronic Publishing*. 33 (4). pp. 313–333.

_____. 2018. Public-Private Partnerships in TVET: Adapting the Dual System in the United States. *Journal of Vocational Education and Training*. 70 (4): pp. 497–523.

Remington, T. F. and P. Yang. 2020. Public--Private Partnerships for Skill Development in the United States, Russia and China. *Post-Soviet Affairs*. 36 (5-6). pp. 495–514.

Risler, M. 1989. Berufsbildung in China (Technical and Vocational Education and Training in China). In *Mitteilungen des Institut für Asienkunde Hamburg*. Hamburg: Institut für Asienkunde.

Sadler, M. E. 1912. England's Debt to German Education. In J. H. Higginson, ed. *Selections from Michael Sadler*. Liverpool, UK: Dejall and Meyorre International Publishers Ltd. pp. 103-105.

Seng, L. S. 2011. *Case Study on "National Policies Linking TVET with Economic Expansion: Lessons from Singapore"*. Bonn: UNESCO.

Seo, S. 2002. Footsteps of the Korean Vocational Training System: Focusing on the Stories behind the Institutionalization Process. Seoul: The Korean Chamber of Commerce and Industry.

State Council of the People's Republic of China. 2015. The general plan of decoupling trade associations and chambers of Commerce from administrative organs. http://www.gov.cn/zhengce/2015-07/08/content_2894118.htm

Stockmann, R. et al. 2000. *Wirksamkeit Deutscher Berufsbildungszusammenarbeit* (The Effectiveness of German Developmental Cooperation) Wiesbaden: Westdeutscher Verlag.

Sul, D 1992. Rural Exodus and Transformation of Urban Labor Market in Korea, 1960~90. *Rural Society*. 2. pp. 145–190.

Sung, J., J. Turbin, and D. Ashton. 2000. Towards a Framework for the Comparative Analysis of National Systems of Skill Formation. *International Journal of Training and Development*. 4 (1). 8–25.

Thelen, K. A. 2004. How Institutions Evolve: *The Political Economy of Skills in Germany, Britain, the United States, and Japan*. Cambridge, UK: Cambridge University Press.

Thøgersen, S. 1990. *Secondary Education in China after Mao*. Aarhus: Aarhus University Press.

Tilak, J. B. 2001. *Building Human Capital in East Asia: What Others Can Learn*. Washington, DC: World Bank Institute.

Tilak, J. G. 1988. Economics of Vocationalization: A Review of Evidence. *Canadian and International Education*. 17 (1). pp. 45–62.

Todd, R., and M. Dunbar. 2018. *Taking a Whole of Government Approach to Skills Development*. Paris: UNESCO Publishing.

United Nations Educational, Scientific and Cultural Organization (UNESCO). 1962. *Recommendation Concerning Technical and Vocational Education*. Revised 1974 and 2001. Paris.

——. 1989. *Convention on Technical and Vocational Education*. Paris.

——. 2015. *Recommendation Concerning Technical and Vocational Education and Training (TVET)*. Paris.

UNESCO International Bureau of Education. n.d. Glossary: IBE-UNESCO. http://www.ibe.unesco.org/en/glossary-curriculum-terminology/c/competency-based-curriculum.

UNESCO and ILO. 2018. *Taking a Whole of Government Approach to Skills Development*. Geneva.

UNESCO-UNEVOC. 2008. *Vocational Education and Training: A Come-Back to the Development Agenda?* Bonn.

Wagner, H.-G. 1999. Deutsch-Chinesische "Lern-Konflikte" (Sino-German "Conflicts of Learning". *Zeitschrift für Berufs- und Wirlschaftspädagogik*. 95 (3). pp. 344–362.

Watson, K. 1994. Technical and Vocational Education in Developing Countries: Western Paradigms and Comparative Methodology. *Comparative Education*. 30 (2).

Wang, X. 2014. *Social Construction of Skill Formation: Sociological Analysis of Transformation of Apprenticeship*. Beijing: China Social Science Archive Press.

Williams, S., and P. Raggat. 1998. Contextualising Public Policy in Vocational Education and Training: The Origins of Competence-based Vocational Qualifications Policy in the UK. *Journal of Education and Work*. 11. pp. 275–292.

Wilson, D. N. 2001. *The German "Dual System" of Occupational Training: A Much-Replicated but Oft-Failed Transfer*. Paper presented at the annual meeting of the Comparative and International Education Society. San Antonio.

Wilson, D. N., M. Kennedy, and M. De Jocas 1999. *Expansion of TVET in Botswana: Needs Review and Project Justification*. Toronto: Educational Consulting Services Corp.

Winch, C. 2006. Georg Kerschensteiner—Founding the Dual System in Germany. *Oxford Review of Education*. 32 (3): pp. 381–396.

Wollschläger, N. and É. F. Guggenheim. 2004. A History of Vocational Education and Training in Europe—From Divergence to Convergence. *European Journal: Vocational Training*. 32. pp. 1–3.

World Bank. 1991. Vocational and Technical Education and Training. Washington, DC.

_____. 1992. *China: Reforming the Urban Employment and Wage System*. Washington, DC. http://documents.worldbank.org/curated/en/813631468213280848/pdf/multi-page.pdf.

_____. 2011. Learning for All: Investing in People's Knowledge and Skills to Promote Development. Washington, DC. https://openknowledge.worldbank.org/handle/10986/27790.

_____. 2020. *The Human Capital Index 2020 Update: Human Capital in the Time of COVID-19*. Washington, DC.

Xu, C. 2011. The Fundamental Institutions of China's Reforms and Development. *Journal of Economic Literature*. 49 (4). pp. 1076–1151.

Yang, P. 2017. Coordinating Public--Private Partnership in VET Sector: Evidence from China. *Journal of the New Economic Association*. 36 (4). pp. 189–198.

Yoo, S. et al. 1967. *Comprehensive Research on Science and Technology Personnel Related to the Population of Korea*. Ministry of Science and Technology.

Yoon, J. et al. 2012. *Analysis on Development and Achievement of Compulsory Elementary Education after the Korean War*. Seoul: Korea Development Institute.

Young, M. 2011. National Vocational Qualifications in the United Kingdom: Their Origins and Legacy. *Journal of Education and Work*. 24 (3–4). pp. 259–282.

www.ingramcontent.com/pod-product-compliance
Lightning Source LLC
Chambersburg PA
CBHW041247240426
43669CB00028B/2999